DOING STATISTICS

With

Excel™ 97

Software Instruction and Exercise Activity
Supplement

Marilyn K. Pelosi, Ph.D.
Western New England College

Theresa M. Sandifer, Ph.D.
Southern Connecticut State University

Jerzy J. Letkowski, Ph.D.
Western New England College

John Wiley & Sons, Inc.
New York • Chichester • Weinheim • Brisbane • Singapore • Toronto

This book is dedicated to our families.

Table of Contents

Procedures		Page

Chapter 14

Preface

To The Student:

Data! Data! Data! It is everywhere and growing at an incredible rate. As our computer technology allows us to capture, store and retrieve more and more data, we must be able to transform that data into information. The information can then be used to make business decisions at all levels ranging from day-to-day decisions to short-term decisions to long-term strategic planning decisions. We can no longer compete internationally by making decisions based solely on experience and intuition. We must be able to see the information in the data.

This workbook is designed to help you learn how to see the valuable information in data. Each chapter is designed to allow you to explore a single statistical concept by investigating a data set. The situations which led to the collection of the data sets are based on actual consulting experiences of the authors.

It becomes quickly obvious that you can not investigate a data set without the assistance of a computer package. The computer package utilized in this workbook is known as Microsoft Excel 97. The commands necessary to use this package are explained in each chapter.

To The Instructor:

This workbook is designed to be used by students enrolled in a basic course in Statistics. It is built around the premise that students must *Do Statistics* in order to truly understand the statistical concepts. Thus, each chapter reinforces a topic from a standard statistics course. It is expected that the student can work independently or as a member of a team of students.

The workbook is self-contained in that all of the instructions needed to use the software are explained in the appropriate chapter.

Each chapter contains two types of exercises: (1) exercises to help the student learn the command structure of Excel 97and (2) exercises to help the students see the statistical concepts in action. The chapter is motivated by an actual problem drawn from the authors consulting experiences. There is one data set for each chapter which is tied to the motivating problem. Each of the datasets is on the disk which accompanies this workbook. The data set is then explained and the remainder of each chapter is dedicated to two major goals:

(1) to provide the student with detailed command instructions for using Excel 97 to accomplish the desired statistical analysis;

and

(2) to provide the student with a set of structured exercises designed to direct the students statistical thinking in order to reinforce the statistical tool of the chapter.

Statistical cases are explored in Chapters 4 through 15. Chapter 2 provides an introduction to Windows 95. Chapter 3 explains basic Excel 97 commands.

Chapter 1 "Why *Do* Statistics in Excel 97?"

An Introduction to the Workbook

Section 1.1 Overview

This chapter discusses the following topics:

- Objectives of the workbook
- Description of the Data Sets
- Organization of the Workbook
- How to use the Workbook

Section 1.2 Objectives of the Workbook

This workbook has two major objectives. They are:

(1) to allow you to *see* the statistical concepts taught in a standard course in statistics at work by *doing statistics* with real life data sets using the Excel 97 software and;

(2) to help you *learn how to use Excel 97*.

At first glance, these may appear to be rather separate and unrelated objectives. In fact, they are supportive of each other. The workbook is designed around the premise that in order to truly understand the statistical concepts you must *do statistics* using real life data sets and in order to *do statistics* you must use a statistical software package. Thus by exploring the data sets you will learn both the statistical tools and the command structure of the software.

Section 1.3 Description of the Data Sets

Each chapter uses one data set whose subject is hinted at in the title of the chapter. Each data set is as an Excel 97 workbook file, on the disk accompanying the workbook.   In a few cases, a data set may be used for more than one chapter. In these cases, the details of the data set are provided in the first chapter that uses this data set.

The data sets are based on actual business situations and in all cases reflect actual problems. They are drawn from the consulting experiences of the authors. The names of the companies have not been provided in order to maintain confidentiality agreements. However, the problem descriptions, the variables and the data are real. The data sets are also large enough to (1) allow you to see the need for a statistical software package and (2) to explore patterns and relationships.

Some of the data set files contain Excel macros that can assist you in doing some of the more complex operations and data entry chores. There is also one Excel file that contains all the macros. Its name is **MacDoIt.xls** and it is stored in the **Macros** folder. You may wish to open this file, if you want to apply some of the macros to your own data sets.

Section 1.4 Organization of the Workbook

The workbook is divided into two major parts.

Part I: Basics of Windows and Excel 97

Chapter 2 explains some basic Windows commands. Chapter 3 gets you started with Microsoft Excel 97. It covers the basic Excel structure and commands. After reading this chapter you will be ready to move on to the remaining chapters of this workbook.   Specific commands and software features will be explained in the chapters where they are needed for the statistical analysis.

Part II: Statistics

Chapters 4-16 are the chapters where you will be *Doing Statistics*.

Each chapter is designed to allow you to explore a single statistical concept by investigating a data set. The chapters follow the normal sequence of topics in an introductory statistics course. Each chapter has the following structure:

- Summary of Statistical Objectives

- Problem Statement
  The problem statement section of each chapter describes the business situation which led to the collection of the data.

- Characteristics of the data set
  This section of each chapter explains the details of the particular data set including information such as the filename, the name and column location of the variables, size of the file, etc.

- Tutorials
  The next few sections of each chapter give the instructions for how to use Excel to accomplish the particular statistical concept. Examples are provided using the data set for that chapter.

- Investigative Exercises
  These exercises are designed to allow you to analyze the data set using the specific tools explained in the chapter. There is space provided for you to either paste in printed output from Excel or to write in the answers by hand based on the screen display from Excel.

  The exercises are highly structured in the early chapters leading you to a directed analysis of the data set. You will always be asked to draw conclusions and make recommendations on the basis of your analysis.
  As the chapters progress, it is expected that you will become increasingly familiar with the sorts of questions that should be

asked, the types of calculations which might be informative and the types of comparisons which should be examined.

Section 1.5 How to Use the Workbook

This workbook is designed to be used by the students taking an introductory course in statistics. It is assumed that the student has been presented the traditional lecture material for each of the concepts prior to working with the corresponding chapter of the workbook.

The workbook chapters could be assigned as homework on an individual basis or to teams of students. Alternatively, the workbook chapters could be used during class lab sessions.

Section 1.6 Web Pages

For more information and extra materials related to this workbook, please visit the following web pages:

 http://www.wiley.com/college

 http://mars.wnec.edu/doingstats

Chapter 2 The Basics of Windows

Section 2.1 Overview

This chapter will review some of the basic features of the Microsoft Windows 95 Operating System, herafter referred to as Windows 95 or just Windows. It is not designed to be a comprehensive introduction to Windows. It only covers those features of Windows that are essential for using Microsoft Excel 97 and this workbook. It assumes that your are familiar with basic keyboard and mouse operations and that Windows 95 is installed and **currently running** on the computer you are using.

Section 2.2 Starting Windows 95

Unlike Windows 3.1, Windows 95 is a standalone operating system. Therefore, starting Windows 95 requires no more than turning your computer on. If your computer is connected to a network, you may be asked to select some options from one or more menus and/or a password may be required. Ask your system administrator or instructor for directions, if necessary.

Section 2.3 The Windows 95 Desktop

Windows 95 is a hierarchical system of objects. The highest level-object in its structure is the *desktop*, which is what you see when you start a Windows 95 based computer. Figure 2.1 shows an example of the Windows 95 desktop. This is a object that can contain directly or indirectly other Windows objects (programs, documents). All the objects are shown on the desktop as *icons*. If you double-click an icon, the system will open the window represented by the icon. If you right-click an icon or other Windows object, you will see a shortcut menu associated with the object. The menu contains common commands that can be applied to the object. For example, by right-clicking the **My Computer** icon, you will be able to open or explore it, or to check its properties, or to find a file, or to do other tasks.

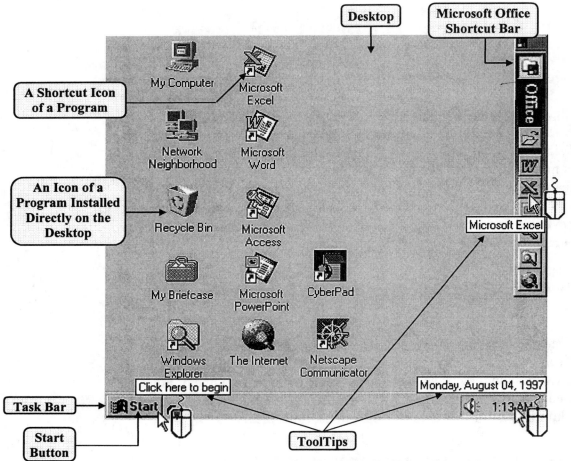

Figure 2.1 The Windows 95 Desktop

The desktop is a workspace used for organizing programs, documents, and other objects, as well as for runnig programs as windows. Depending on the system you use, the desktop may not look exactly as the one shown above. It may have different program icons and shortcuts. Nontheless, all the objects are used in a common way.

An important part of the desktop is the *task bar*. It contains buttons and/or icons of rinning programs. The task bar shown in Figure 2.1 contains buttons of three programs (**Start**, **Volume Control**, and **Date/Time**), all inserted onto the task bar upon the start of Windows 95. Notice that the **Date/Time** button shows only the current time. In order to see the date, you must place the mouse pointer on the time, as shown in Figure 2.1.

The most important **Task Bar** button is the **Start** button. It can be used to start especially those programs that do not have their icons (shortcuts) on the desktop. In a way, the **Start** button is the beginning of the Windows menu structure. If you click the **Start** button using the <u>left</u> mouse button, Windows will open the pull-up menu. On the other hand, if you click **Start** using the <u>right</u> mouse button, you will be able to open the menu as a window.

Section 2.4 Using the Mouse

One of the important features of the **Graphical User Interface (GUI)** operating systems is their use of a pointing device like a mouse, track ball, touch pad, etc. Once you know how to operate the mouse, you will have no problems with the use of the other pointing devices.

Although most of the Windows procedures can be performed using the keyboard, many of them can be done faster, more precisely, conveniently, and otherwise more efficiently with a mouse. Pointing, clicking, double-clicking, drag-and-dropping are essential buzzwords of today's GUI jargon. They represent typical mouse operations.

With a two-button mouse, for example, with a Microsoft mouse, you can perform the following operations:

Table 2.1 Standard Mouse Operations in Windows 95

	Pointing (Mouse Up)	Place the mouse pointer on an object without pressing any button. Some Windows objects triger some action when you just move/place the mouse pointer over them. In many instructions, this operation is also shown as the last step of a mouse dragging process.
	Left or Right Mouse Button Down	Place the mouse pointer on an object and hold down the left or right button (as indicated). In many instructions, this operation is also shown as the first step of a mouse dragging process.

Note:

In order to start a program, open a document, or access some system services, click the **Start** *button and follow the menu options.*

	Clicking the Left Button	This is a common mouse operation. Its job is usually to select Windows objects. To click means to press and right away release the button.
	Clicking the Right Button	In many applications, this operation is used to reveal properties of Windows objects pointed at by the mouse pointer or to pop up a shortcut (context sensitive) menu.
	Double-clicking the Left Button	This operation is designated to both selecting and activating (launching) Windows objects. It involves two quick clicks of the mouse button.
	Dragging	This operation is used usually to move, copy, expand, combine, or resize Windows objects. It requires that you first point to an object and then press and hold down the mouse button while moving the mouse. The arrows "attached" to the mouse images shown on the left indicate the dragging direction (here — to the right). A dragging operation may involve the left or right mouse button being held down. This operation is frequently called a drag'n drop operation. In many instuctions, the entire drag and drop process is illustrated by the following mouse states: **Mouse Down** **Dragging** **Mouse Up**

Note: In this workbook, unless otherwise indicated, a **click** operation expects clicking the **left** mouse button, and the **drag'n drop** or **drag** operation also expects the **left** button being held down. The symbolic mouse images shown above will be used to visually describe mouse operations. In all step-by-step instructions, the mouse images may also contain step numbers. In the example shown on the left-hand side, the third step of some instruction is to click the **Save** button.

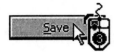

Depending on both the object pointed at and the state of a running program, the program may show different shapes of the mouse pointer. By changing the mouse pointer's shape, Windows programs and the operating system show what kind of mouse operation can be done with the object pointed at by the mouse pointer. Figure 2.2 shows standard Windows 95 mouse pointers.

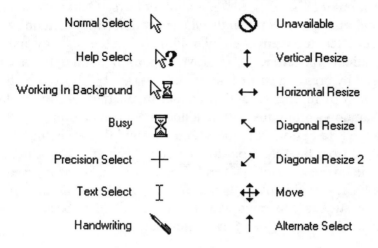

Normal Select		Unavailable	
Help Select		Vertical Resize	
Working In Background		Horizontal Resize	
Busy		Diagonal Resize 1	
Precision Select		Diagonal Resize 2	
Text Select		Move	
Handwriting		Alternate Select	

Figure 2.2 Typical Windows 95 Mouse Pointers

Section 2.5 Using the Keyboard

The Windows 95 operating system and most of its application programs support a common way of performing typical tasks via the keyboard, mouse, and both keyboard and mouse. Many of the keyboard/mouse operations have been inherited from the previous version of Windows. A few new operations have been added.

In Windows 95, virtually all the mouse operations can be performed using the keyboard. They can be done either directly through keyboard shortcuts or by simulating the mouse actions (move, click, drag, etc.). For example, rather than clicking the **Start** button, one can press the CTRL+ESC key combination. In order to be able to simulate the mouse using the keyboard, the **MouseKey** feature must be activated. Check the last part of this section for more details about this feature.

Note:
In order to do CTRL+ESC, *hit* ESC *while* CTRL *is being held down.*

In many situations, especially when working with text, using the keyboard is a more convenient way to execute editing and formatting commands. Also some window manipulations and even drawings can be done more precisely using the keyboard rather than mouse. Table 2.2 summarizes the most popular keyboard shortcuts used in Windows 95. Notice that each application program may introduce more specialized keyboard shortcuts. They usually involve a combination of ⌊ALT⌋, ⌊CTRL⌋, ⌊SHIFT⌋, and other keys. Generally, menu options and dialog box commands whose caption (title) contains an underlined letter can be invoked by holding down the ⌊ALT⌋ key and pressing quickly the **letter**. Drop-down or pop-up menu options can be accessed by pressing vertical arrow keys (⌊↑⌋, ⌊↓⌋) followed by the ⌊ENTER⌋ key, or by hitting the option's underlined letter. For example, in most Windows programs, saving a document can be done by invoking the **File** option from the **Menu Bar** and then selecting the **Save** option from the **File** drop-down menu. Using the keyboard, this can be done by pressing ⌊ALT⌋+⌊F⌋ and ⌊S⌋ (first the ⌊F⌋ key is pressed while the ⌊ALT⌋ key held down, then the ⌊S⌋ key is pressed alone). A series of keystrokes, separated with a comma will describe such commands. Here is the **File** | **Save** example:

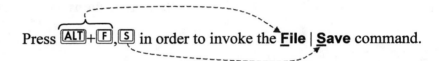

Press ⌊ALT⌋+⌊F⌋,⌊S⌋ in order to invoke the **File** | **Save** command.

Notice that the keystrokes are separated with a comma and the command captions — with a vertical bar.

Table 2.2 Basic Windows Keyboard Commands

Keystroke	Action
⌊CTRL⌋+⌊ESC⌋	Pop up the **Start** menu.
⌊CTRL⌋+⌊ALT⌋+⌊DELETE⌋	Invoke the **Close Program** dialog box.
⌊F1⌋	Get help on a selected dialog box item or invoke the help system.
⌊F2⌋	Rename an item.
⌊F3⌋	Invoke the **Find** dialog box.
⌊ALT⌋+⌊TAB⌋	Switch to the previous program window. Or switch to another window by holding ⌊ALT⌋ while repeatedly pressing ⌊TAB⌋ until the window is selected.
⌊ALT⌋+⌊F4⌋	Quit a Windows program.
⌊↑⌋,⌊←⌋,⌊↓⌋,⌊→⌋	Move between shortcut icons on the desktop and between menu options.

SHIFT + **F10**	Pop up the shortcut menu of a selected item.
ALT + **ENTER**	View the properties of an item.
SHIFT	To bypass some automatic procedures. For example, when a compact disk with a **Setup** program is inserted, Windows automatically starts the program. To prevent this behavior, insert the CD-ROM while holding the **SHIFT** key.
M	Delete an item.
SHIFT + **DELETE**	Delete an item without placing it in the **Recycle Bin**.
CTRL + **X**	Cut a selected object to the **Clipboard**.
CTRL + **C**	Copy a selected object to the **Clipboard**.
CTRL + **V**	Paste the object from the **Clipboard**.
CTRL + **Z**	Undo an operation.

As mentioned, the keyboard can also be used to simulate basic mouse operations. This feature can be set in the **Accessibility Properties** dialog box which is located in the **Control Panel** window. Using the keyboard, do the following steps:

Procedure 2.1 Setting the MouseKeys Feature

	Menu Option / Command	Keystroke
❶	**Start**	**CTRL** + **ESC**
❷	**Settings**	**S**
❸	**Control Panel**	**C**
❹	**Accessibility Options**	**A**
❺	**Mouse**	**CTRL** + **TAB** Press repeatedly until the **Mouse** tabbed section is revealed (Figure 2.3).

Windows opens the **Accessibility Properties** dialog box.

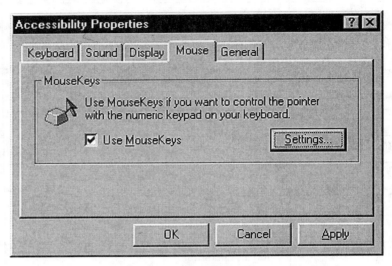

Figure 2.3 The Accessibility Properties dialog box.

Continue the steps of Procedure 2.1:

	Menu Option / Command	Keystroke
❻	Use **M**ouseKeys	[ALT]+[M]
❼	OK	[ENTER]

Windows displays a mouse icon in the message area of the **Task** bar.

Now, with [Num Lock] switched on, you can use the numeric keypad to simulate mouse operations. Table 2.3 shows the keystrokes to simulate the mouse operations.

Table 2.3 MouseKeys

Mouse Operation			Keystroke		
Navigation	↖↑↗ ← → ↙↓↘	*Note:* Hold down **Ctrl** to speed up or **Shift** to slow down.	[7 Home] [↑] [9 PgUp] [←] [→] [1 End] [↓] [3 PgDn]		
Set the Left Click (Default)			[/]		

Set the Both Click	⊡ *
Set the Right Click	⊡ –
Click	⊡ 5
Double Click	⊡ +
Turn the Drag Operation On	⊡ 0 Ins
Turn the Drag Operation Off (Drop)	⊡ . Del

Windows 95 provides a way to customize the **MouseKeys** feature.
Selecting the **Settings** command button, in the **Mouse** section of the
Accessibility Properties dialog box, will open **the Settings for
MouseKeys** dialog box (Figure 2.4). Here, you can set the **Top** speed
and **Acceleration** of the mouse pointer. If the **Use shortcut** option is
checked, you will later be able to turn **MouseKeys on** and **off** by
holding down the left ⒜ALT⒜ and ⒮SHIFT⒮ keys and then pressing the
key ⌨Num Lock⌨. You can also decide whether or not to use the **Ctrl** and **Shift**
key to speed up or slow down the mouse pointer movement. You can
also set the state of the numeric keypad (⌨Num Lock⌨ **on** or **off**) when the
MouseKeys operations are available. Finally, you can instruct
Windows 95 whether or not to show the **MouseKeys** status in the
message area of the **Task** bar. There are three states of this status:

🖱 Left,   🖱 Right, or   🖱 Both.

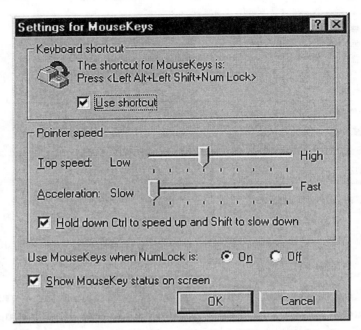

Figure 2.4 The Settings for MouseKeys dialog box.

The **Accessibility** options provide other ways to customize the user interface for those with mobility, hearing, or visual disabilities.

Section 2.6 Windows Structure and Behavior

A rectangle with built-in functional components makes up a window. It is contains many *sensitive elements* that are designed to respond to the mouse and keyboard operations. Figure 2.5 demonstrates basic elements of the window structure organized around the window's frame. Among the window elements shown in this figure, only the ***Control*** box can be accessed directly from the keyboard. All other elements require the mouse.

The window's area is a large *hot zone*. Pointing and clicking anywhere within the area makes the window a *current* (or *active*) window. In the current window, the ***Title*** bar is exposed. By default, the title bar of active windows is blue and the title bar of inactive windows is gray. Some windows, referred to as *Parent* windows, may contain other windows (*Children*). Figure 2.5 shows the Excel program as a parent window and the **Book1** workbook as a child window.

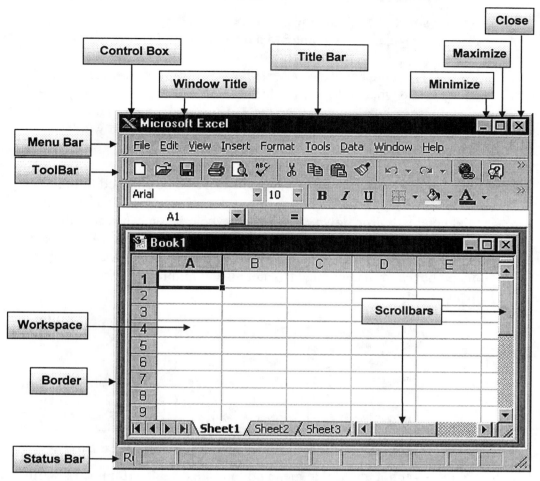

Figure 2.5 Basic Window Components.

Many windows contain three buttons in their upper right corner: *Minimize* (⬛), *Maximize* (⬜) or *Restore* (⬛), and *Close* (⬛) buttons. If you click the **Minimize** button, the window will be reduced to its smallest form referred to as an icon. Such an icon is placed on the **Task** bar. The same operation with the **Maximize** button will blow up the window to the size of the parent window or the screen. After the window has been maximized, the **Maximize** button changes to the **Restore** button. Clicking the latter restores the most recent size of the window. Finally, if you click the **Close** button, the window will be closed and the program running in it will be terminated.

Dragging one of the window *borders* or *corners* can change the vertical and horizontal dimensions of a restored window. In order to change both vertical and horizontal size, place the mouse pointer at one of the corners and drag it in the desired direction. Note that you should not start dragging the mouse before the mouse pointer changes from the regular arrow pointer to a diagonal double arrow pointer (↘).

The **Title** bar is also a sensitive window's zone. It can be used to move the window to another location on the screen. Simply place the mouse pointer anywhere on the **Title** bar and drag it to the new location. If you double click the **Title** bar of a maximized window, the window will be restored and vice versa.

The size and state of a window can also be changed through the **Control** menu. Just click the **Control** box (upper-left corner) and the menu will drop down. Using the keyboard, you can access the menu by pressing ⌘ALT+⎵ (**Alt+Space Bar**) for a Parent window and ⌘ALT+⊟ (**Alt+Dash**) for a Child window. Figure 2.6 shows **Control** menu boxes for a maximized, restored, and minimized window, respectively.

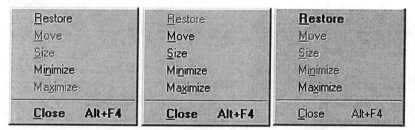

Figure 2.6 Different States of the Control Menu Box.

Notice that <u>not applicable</u> menu options are disabled (dimmed). With the **Move** or **Size** command, you can use the arrow keys to change the position or size of the window.

Figure 2.5 shows an Excel window equipped with a menu system. It is a common Windows feature to implement the main menu as a top *Menu* bar. Figure 2.7 shows more details about the menu system in Excel. In fact, a similar structure and the same conventions can be found in other Windows programs, equipped with a menu system.

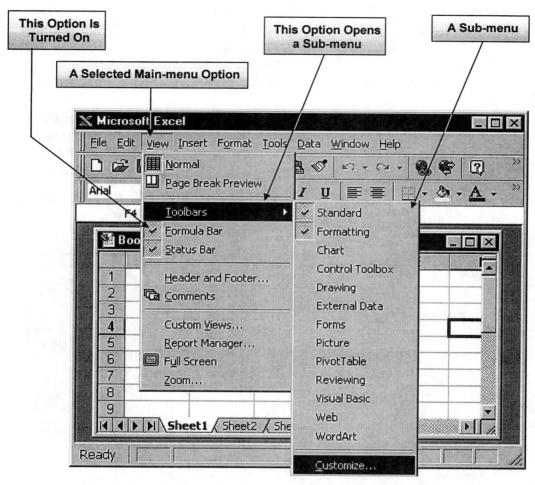

Figure 2.7 A Fragment of the Microsoft Excel Menu System.

When a **Menu** bar option is selected, it activates a ***Drop-down*** menu, which contains its own set of options (commands). A **Drop-down** menu may contain both simple and complex commands. Upon selection, a simple command is carried out immediately, whereas a complex one pops up another menu or a special window referred to as a *dialog box*. Note that the complex commands are followed by an ellipsis (···) or by an arrowhead (►). Some menu options act as switches. They are used to turn some features or properties **on** or **off** (for example: the options **Formula Bar**, **Status Bar**, **Standard**, and **Formatting** as shown in Figure 2.7).

When you execute a complex command that is followed by an ellipsis ($\cdots$), it opens a <u>fixed</u> size window called a *dialog box*. Such a window enables the user to properly customize the command. It makes it possible for the user to construct and implement very sophisticated commands without having to remember their syntax.

A dialog box can be equipped with many interactive and user-friendly input tools. Figure 2.8 shows the **Format Cells** dialog box invoked in Excel by the **Format | Cells...** command.

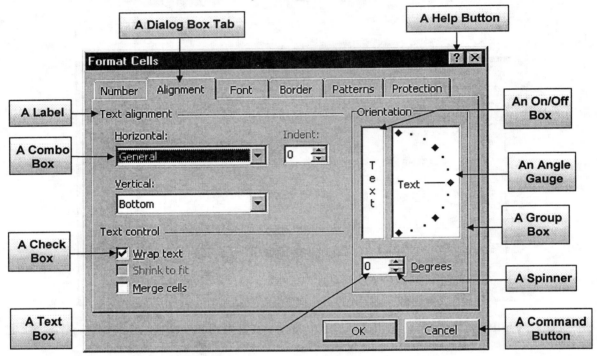

Figure 2.8 The Format Cells Dialog Box.

This dialog box contains seven typical Windows controls:
- *Labels*,
- *Text* box,
- *Check* box,
- *Combo* (*Drop-down list*) box,
- *Command* button,
- *Spinner*,
- *Group* box,
- *Help* button,

and two specialized controls:
- *On/Off* box,
- Angle *gauge* (integrated with a text box and spinner).

The simplest way to select a control is to click it with the mouse. Using the keyboard you can apply the shortcut keys. For example, in the **Format Cells** dialog box, you can select the option **Degrees** by pressing ⌨️ALT+⌨️D. You can also arrive at this option by pressing repeatedly the ⌨️TAB and arrow keys (⌨️←⌨️→⌨️↑⌨️↓). Usually, the **OK** button is associated with the ⌨️ENTER key and the **Cancel** button with the ⌨️ESC key.

The **Format Cells** dialog box, shown above, is a **Tabbed** dialog box. It combines 6 formatting commands: **Number**, **Alignment**, **Font**, **Border**, **Patterns**, and **Protection**). Each of these commands is accessed by clicking its **Tab**. You can also use the ⌨️CTRL+⌨️TAB keyboard command to move from one command page to another.

Labels are used in dialog boxes to annotate other controls, or to display some useful information. The user cannot directly access labels.

Text boxes are powerful and versatile data entry tools. When you place the mouse pointer within a **Text** box, it changes from the regular arrow ➤ to the so-called *I-beam* pointer Ⅰ. When you click the text box, the I-beam changes to a flashing vertical bar | referred to as an *insertion point* or *cursor*. Then you can type and/or edit your text. Use ⌨️DEL or ⌨️BACKSPACE to delete characters and the ⌨️← ⌨️→ ⌨️HOME ⌨️END keys to move the cursor within the box. A **Text** box usually accepts the **Cut/Copy/Paste** commands by means of the respective keyboard commands: ⌨️CTRL+⌨️X, ⌨️CTRL+⌨️C, and ⌨️CTRL+⌨️V. You can select a text by dragging the mouse over it, or by holding down ⌨️SHIFT and using the text navigation keys (⌨️← ⌨️→ ⌨️HOME ⌨️END).

Check boxes are either empty or marked with **X** or ✓. They represent non-exclusive (free) options. Thus, you can mark as many of the **Check** boxes as you need. Click a box or press ⌨️ALT along with the box caption's underlined letter, in order to check the box **on** or **off**. You can also use ⌨️TAB to access the box and then press **Space Bar** to check it **on** or **off**.

A **Combo (Drop-down List)** box is a space-conserving implementation of a regular **List** box. When closed, it looks like a **Text** box with a **Drop-down** arrow ▾ attached at the right side. A closed **Drop-down List** box shows the default or most recently selected option. The **Drop-down** arrow suggests that there are some other alternatives available. To select an option from a **Drop-down List**, first open the list. Using the mouse, simply click on the **Drop-down** arrow ▾. The keyboard way is a bit more complex. First, to get to the box, hold down ⒜ⓁⓉ and press the list caption's underlined letter or repeatedly press ⓉⒶⒷ until the box is selected. If needed, press ⒜ⓁⓉ+⬇, to open the box. When the list is expanded, click the desired option or use the vertical arrow keys ⬇⬆ to move the selection highlight to the wanted option and then press ⒺⓃⓉⒺⓇ (or ⒜ⓁⓉ+⬆ or ⒜ⓁⓉ+⬇).

Command buttons are dynamic components of a dialog box. They represent simple or complex commands. A simple command carries out some task immediately. A **Command** button, having a name followed by an *ellipsis* (···) or by a *double greater-than* sign (≫), represents a complex command. When activated, the former pops up a new dialog box, the latter expands the current dialog box. The **OK** button (command) triggers the process assigned to its dialog box, whereas the **Cancel** command abandons it.

A **Spinner** or **Up/Down** control usually accompanies a **Text** box. It is used to increment or decrement the numeric value stored in the **Text** box. Click the upper arrowhead in order to increase the value, or click the lower arrowhead to decrease the value.

Group boxes are used to integrate other controls. They act as containers. The **Orientation** group box of the **Format Cells** dialog box integrates four different controls: angle **gauge**, **text** box, **spinner**, and **on/off** box. The latter one is used to change the text orientation from horizontal to vertical and vice versa, the former ones are alternative ways for rotating cell contents.

Option (Radio) buttons, shown in Figure 2.9, represent exclusive selections or settings. This is why, in a group of **Option** buttons, only one button can be set on. Click an **Option** button or press [ALT] along with the button caption's underlined letter, in order to check the button **on** or **off**.

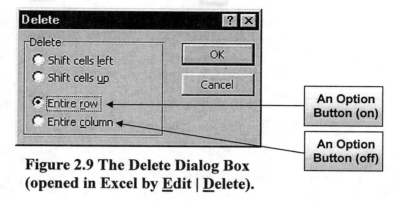

Figure 2.9 The Delete Dialog Box (opened in Excel by Edit | Delete).

If you click the **Help** button, Windows will change the mouse pointer to the **Help Select** pointer. Using this pointer, click the object you want information about and a pop-up message window will tell you more about the object. Figure 2.10 demonstrates getting help on **Wrap Text** in the **Format Cells** dialog box.

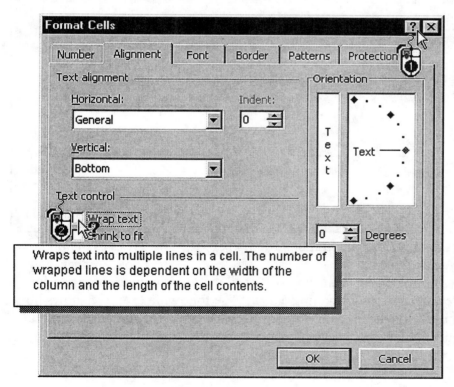

Figure 2.10 Using the Help Pointer in a Dialog Box.

Interestingly, in a dialog box, when you right-click a control, the dialog box will show the **What's This?** button. Subsequently, clicking the button will open the quick a help box like the one shown in Figure 2.10.

Section 2.7 My Computer

The **My Computer** program provides an easy access to computer resources and files. Its icon is usually located in the upper-left corner of the **Desktop**. Figure 2.11 shows an example of the **My Computer** window. From this window, you can get to your computer drives, printers, network, and other resources. The **Control Panel** program gives you handy tools for managing your computer's hardware and software. Using the **Dial-Up Networking** program, you can set up a connection to the Internet.

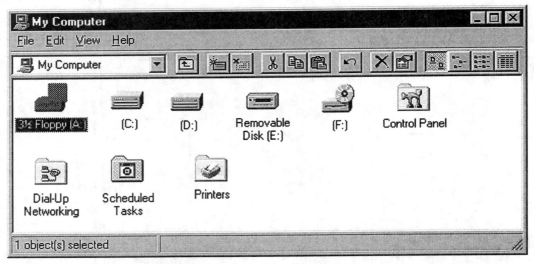

Figure 2.11 The My Computer Program Window.

Common tasks, you will need the **My Computer** program for, are copying and formatting a disk. The following example shows how to create a backup copy of a **3½** floppy disk. Formatting a diskette is quite similar (in the step ❸ of Procedure 2.2, select the option **Format...** instead of **Copy Disk...**).

Procedure 2.2 Duplicating a Floppy Disk.

	Task	Mouse
❶	Insert the floppy disk to be duplicated into the drive and start the **My Computer** program.	Double-click the **My Computer** icon. My Computer
❷	Open the **3½ Floppy** shortcut menu.	Right-click the **My Computer** icon. 3½ Floppy [A:]
❸	From the menu, select the **Copy Disk...** option.	Click (left-click) the **Copy Disk...** option. **Open** Explore Find... Copy Disk... Format... Paste Create Shortcut Properties
❹	In the **Copy Disk** dialog box, select the **Copy from: 3½ Floppy option.**	In the Copy from panel, click the **3½ Floppy** item. Copy from: 3½ Floppy [A:]
❺	**Start** the disk copy process.	Click the **Start** command button. Start
❻	When the message box ⓘ pops up, remove the source disk, insert the destination disk, and **okay** this operation.	Click the **OK** command button. OK
❼	When done, close the **Copy Disk** dialog box.	Click the **Close** command button. Close

Section 2.8 Windows Explorer

Using the **My Computer** program, we can explore and manage Windows objects in a single-window view. *Windows Explorer* provides a more sophisticated environment for program, disk, and file management. **Windows Explorer** provides a tree-folder-structure and contents view of the computer system and rich functionality for manipulating and managing disks, files, and programs. Figure 2.12 shows a *collapsed* view of **Window Explorer**. The **Desktop** folder represents the entire computer. The **All Folders** pane shows the

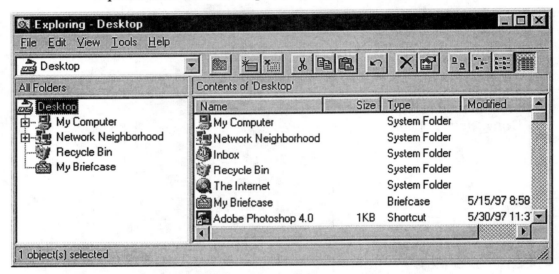

Figure 2.12 The Desktop Tree-Structure in Windows Explorer.

Note:
*Clicking the **Plus** box opens but does not select the folder. Double-clicking a folder **both** expands and selects the folder.*

folders (sub-folders) branching out from the **Desktop** folder. The **Contents** pane includes objects (folders and files) that belong to the folder selected in the **All Folders** pane. With the **Desktop** folder is selected, the **Contents** pane displays all the folder and shortcut icons that we can be seen on the desktop screen. The **Plus-sign** box in front of a folder indicates that the folder contains some objects (other folders or files). If you click the box, the folder will expand and that plus sign will change to a minus sign. On the other hand, if you click the **Minus-sign box**, its folder will close itself (collapse). To find out more about how the folder structure can be explored, try the click, double-click, and right-click operations on the folders.

The following instruction shows how to use Explorer to open the **Readme.txt** file, stored in the **Ch02** folder on the workbook diskette.

Procedure 2.3 Opening the Readme.txt File in Explorer.

	Task	Mouse
❶	Insert the workbook floppy disk in the drive and click the **Start** button.	
❷	From the pull-up menu, select the option **Programs**.	
❸	From the **Programs** sub-menu, select the **Windows Explorer** option.	
❹	Unless the **My Computer** folder is already expanded, click the **Plus (+)** box in front of it.	
❺	Expand the **3½ Floppy** folder.	
❻	Select the **Ch02** folder.	
❼	Open the **Readme.txt** document.	

Windows Explorer opens the **Readme.txt** document in a **Notepad** window as shown in Figure 2.13. In order to close the window, click its upper-right-corner **Close** button.

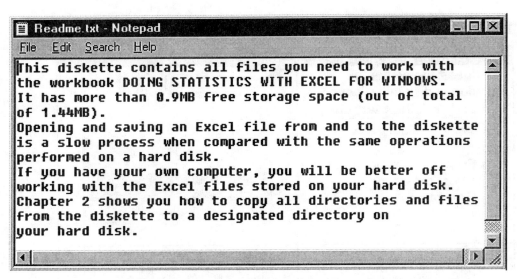

Figure 2.13 The Readme.txt Document in a Notepad Window.

Note:
Not registered documents have this icon:

Document files of all properly registered programs can be opened directly from **Windows Explorer**. Names of such documents are accompanied by some meaningful icons, for example: ▣ (**Word**), ▣ (**Excel**), ▤ (**Notepad**).

If you have your own computer, you can increase your productivity by working with files stored on your computer's hard disk. The following instruction shows how to copy all folders and files from the workbook data diskette onto a new folder, **DoingStats**, on drive **C:**. First you will create the folder and then you will use the drag-and-drop operation to copy the contents of the diskette.

Procedure 2.4   Copying Data Files to a Hard Drive

	Task	Mouse
❶	If your **Explorer** is still open and it shows the folder tree of the data disk, close the **3½ Floppy** folder. Otherwise, in order to start **Explorer** and open the **MyComputer** folder (if needed), perform steps ❶ - ❹ of Procedure 2.3.	Desktop, My Computer, 3½ Floppy (A:), Ch02, Ch03
❷	Select the **C:** drive.	Desktop, My Computer, 3½ Floppy (A:), (C:)
❸	Invoke the **File | New | Folder** command.	Exploring - C:\ File Edit View Tools Help New ▶ Folder, Shortcut, Create Shortcut
❹	Right away, type **DoingStats** and press ⌈ENTER⌋.	DoingStats
❺	Expand the **C:** folder.	Desktop, My Computer, 3½ Floppy (A:), (C:)
❻	Select the **3½ Floppy** folder.	Desktop, My Computer, 3½ Floppy (A:)
❼	In the **Contents of 'A:\'** folder pane, select all folders **Ch02** through **Ch15** (❶ click folder **Ch02**, ❷ hold down ⌈SHIFT⌋, and click folder **Macros**).	Contents of 'A:\' Ch02 Ch08 Ch14 Ch03 Ch09 Ch15 Ch04 Ch10 Macros Ch05 Ch11 Ch06 Ch12 [Shift] Ch07 Ch13

Note:
Click the **List Toolbar** *button, in order to have a similar view type in the* **Contents** *pane:*

List

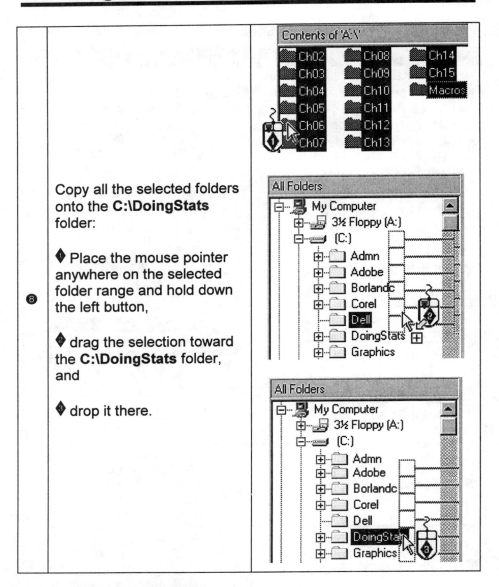

❽ Copy all the selected folders onto the **C:\DoingStats** folder:

◆ Place the mouse pointer anywhere on the selected folder range and hold down the left button,

◆ drag the selection toward the **C:\DoingStats** folder, and

◆ drop it there.

Explorer shows the animated **Copying...** message box. If you click the **C:\DoingStats** folder, you will see all folders **Ch02** through **Ch15** and **Macros** in the **Contents of 'C:\DoingStats'** folder pane.

Section 2.9 Exiting Explorer and Windows

In order to close Windows Explorer window, double click its **Control Menu** box or click the **Close** button.

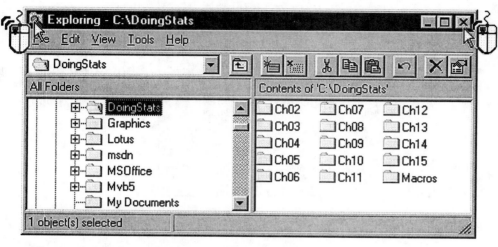

Figure 2.14 Closing Windows Explorer.

Closing a window is a common Windows operation. In the same way, you can close most of the document and program windows. Some dialog boxes may not be equipped with a **Control Menu** box and/or the **Close** button. In such situations, look for an **Exit**, **Quit**, or other window terminating command button.

<u>**Always**</u>, before turning your computer off, shut down Windows 95!

❶ Click the **Start** button and
❷ from the pull up menu, select the **Sh**u̲**t Down** option.

Windows opens the **Shut Down Windows** dialog box.

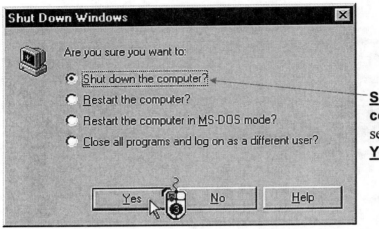

With the **Shut down the computer?** option selected, click the **Yes** button.

Chapter 3 The Basics in Microsoft Excel 97

Section 3.1 Overview

This chapter covers the basic structure and commands of Microsoft Excel 97 for Windows 95. After reading this chapter, you will be ready to move on to the remaining chapters of this workbook. Specific commands and software features will be explained in the chapters where they are needed for the statistical analysis.

There are some more advanced statistical features of the software that are not covered in any of the chapters in this workbook since they are not typically covered in an introductory statistics course. After you have completed the workbook, you should be familiar enough with Excel that you can use the features if you know the statistical theory. Excel has an excellent on-line help system to provide you with more specific information about commands and functions related to the statistical analysis.

Section 3.2 Launching Excel

There are many ways to launch Excel. Depending on how your computer is configured, you can launch Excel via:
- Microsoft Office (MSO) shortcut bar,
- Shortcut to Excel icon on the Desktop,
- Start button,
- Microsoft Explorer,
- My Computer.

Note:
Explore Chapter 2 for more information about the **Start** *button,* **Explorer,** *and* **My Computer**.

The MSO shortcut bar is usually located along the top or right edge of the screen. Click the Microsoft Excel button and MSO will

launch the program with a new Excel document file (a workbook named **Book1**).

Note:
Check if the **New** *and* **Open Office Document** *options are also available from the* **Start** *menu.*

If there is no Excel button, you can start the Excel program with a new workbook document by first clicking the **New Office**

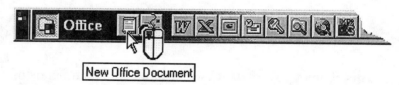

Document button and then double-clicking the **Blank Workbook** icon

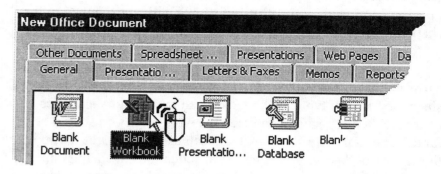

Trouble:
If double clicking doesn't work, click the icon and then also click the **OK** *button.*

in the **New Office Document** dialog box. MSO will start Excel and open a new workbook (**Book1**).

If you intend to work with an existing Excel workbook, previously saved on a disk, your you may wish to click the **Open**

Office Document button and then complete the **Open Office Document** dialog box.

Suppose that you would like to work with the Chapter 4 data file. You would need to open the **Ch04Dat.xls** file, an Excel workbook. The following procedure shows how to do it.

Procedure 3.1   Opening an Excel Workbook from the Open Office Document Dialog Box

	Task Description	Mouse
❶	On the MSO shortcut bar, click the **Open Office Document** button or ◆ click the **Start** button and ◆ select the **Open Office Document** option.	*or* Open Office Document
❷	In the **Open Office Document** dialog box, click the drop-down arrow at the **Look in:** box.	Look in: ☐ My Documents
❸	From the **Look in:** list, select the **3½ Floppy** option.	☐ My Documents Desktop My Computer 3½ Floppy (A:) (C:) My Documents
❹	Open the **Ch04** folder (double-click it).	☐Ch02 ☐Ch12 ☐Ch03 ☐Ch04
❺	Double-click the **Ch04Dat.xls** file or ◆ click the file and ◆ also click the **Open** button.	**Open Office Document** Look in: ☐ Ch04 Ch04Dat.xls *or* Ch04Dat.xls Open

MSO launches Excel, which in turn opens the **Ch04Dat.xls** workbook. Figure 3.1 shows the program (parent-Excel) and document (child-workbook) windows along with interpretation of typical components of the Excel window.

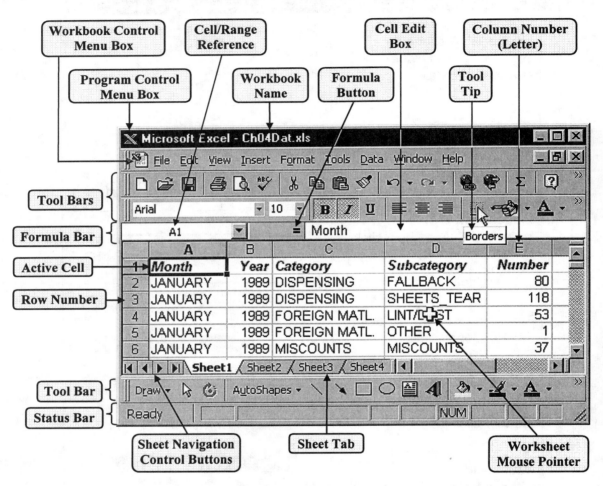

Figure 3.1 Excel Program Window with an Open Workbook

Note:
For interpretation of other window components refer to Figure 2.5.

An Excel 97 workbook is a collection of sheets (worksheets) each made of **256** columns and **65,536** rows.

Worksheet cells are designed to store numbers, text, and formulas. A worksheet can also store graphical objects (charts, pictures and drawings) and other Windows objects.

Entering a number or text into a cell is an easy task. First click the cell or move to it using the navigation keys. The cell becomes then the *active cell*. Next, type the number or text and press ENTER, CTRL+ENTER, or click the *check (enter) button* (☑) on the formula bar. Notice that if you press ENTER, the *cell pointer* will move down to the next row (the same effect has the ⏎ key). Use one of the other arrow keys (⬆, ⬅, ➡) to move the cell pointer in the up, left, or right direction.

Using formulas, you can perform a variety of calculations and other data processing operations. In Excel, a formula always starts with an equal sign (=). It may contain constant values, cell or range references, operators (+,-,*,/,^,%,&), and worksheet functions. In the following chapters, you will see many examples of Excel formulas.

Before going to the next section, close the **Ch04Dat.xls** workbook. Do not save this workbook, if asked. Procedure 3.2 shows alternative ways to close the workbook. Note that closing documents in other Windows programs (Lotus 1-2-3, Quattro Pro, PowerPoint, etc.) can be done in about the same way.

Procedure 3.2 Closing a Workbook Window

Task Description	Mouse/Keyboard
❶ Click the workbook window's **Close** button or double-click the workbook window's **Control Menu** box or execute the **File | Close** command or press CTRL+W.	

Note:
Try the following navigation key to find out what they do:

⬆ ⬅ ⬇ ➡
HOME END
TAB SHIFT+TAB
CTRL+HOME
Page Down
Page Up

Formula Operators:
+   *addition*
−   *subtraction*
*   *multiplication*
/   *division*
^   *exponentiation*
%   *percent (division by 100)*
&   *text concatenation (combination)*

Note:
If you want to take a break now and close Excel, click the program window's **Close** *button:*

Section 3.3 Statistical Data in Excel

Although Excel, in general, does not impose any particular restrictions on the structure of an application, when dealing with statistics, it is strongly recommended that data be arranged in columns.

In Excel, each worksheet column has its own name, a letter or a combination of two letters (**A,B,...Z, AA,AB,...AZ, BA,BB, BZ,...**). Nevertheless, in most cases, you will need to supply your own names for each of the columns of the data. Each data column represents one *Variable*, a collection of specific observations to be analyzed using statistical tools. Since there are **65,536** rows in each worksheet, it is feasible to use Excel for variables that do not exceed **65,535** observations.

Figure 3.2 shows an example of data organized in three variable columns. The instruction that follows (Procedure 3.3)

	A	B	C
1	*Date*	*Shift*	*Quantity*
2	2/1/96	1	992
3	2/1/96	2	517
4	2/1/96	3	331
5	2/2/96	1	992
6	2/2/96	2	489
7	2/2/96	3	329

Figure 3.2 Collection of Three Variables

demonstrates how to enter the data. Notice that the data represents some production quantities (*Quantity*) recorded at the end of each of the three shifts (*Shift*) during a period of two days (*Date*).

Procedure 3.3 Entering Data into a Worksheet

Note:
Section 3.2 tells you more about how to launch Excel.

	Task Description	Mouse / Keyboard
❶	If Excel is running, open a new workbook, otherwise launch Excel.	

❷	Type the word **Date** in the cell A1 and press ↵. Type the word **Shift** in **B1** and press ↵. Type the word **Quantity** in **C1** and press CTRL + ENTER.	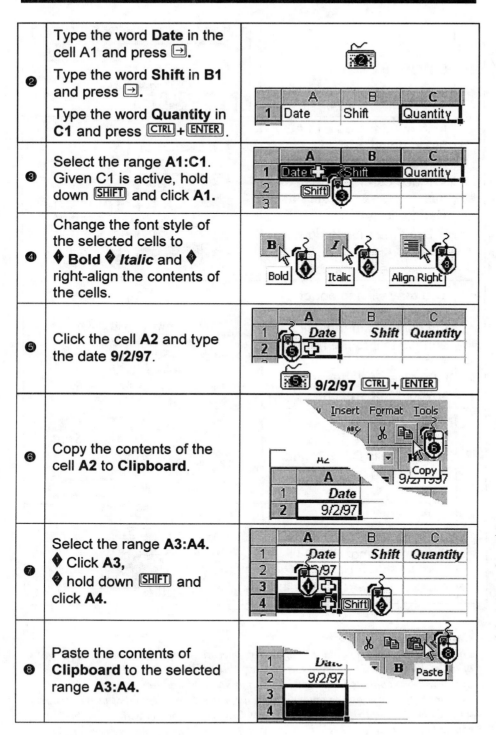
❸	Select the range **A1:C1**. Given C1 is active, hold down SHIFT and click **A1**.	
❹	Change the font style of the selected cells to ◆ **Bold** ◆ *Italic* and ◆ right-align the contents of the cells.	
❺	Click the cell **A2** and type the date **9/2/97**.	
❻	Copy the contents of the cell **A2** to **Clipboard**.	
❼	Select the range **A3:A4**. ◆ Click **A3**, ◆ hold down SHIFT and click **A4**.	
❽	Paste the contents of **Clipboard** to the selected range **A3:A4**.	

Note:
The easiest way to select a range of cells in Excel, is to click the first cell and then SHIFT *click the last cell, or vice versa.*

Note:
The Fill Handle is a small box located in the lower-right corner of the range/cell selection frame:

Fill Handle

By dragging this box, you can copy formulas or create a series of number, dates, or other sequential values.

Note:
CTRL+C *is a shortcut of the* **Edit | Copy** *command (same as* ALT+E,C*). The shortcut for the* **Edit | Paste** *command is* ENTER *or* CTRL+V *(same as* ALT+E,P*).*

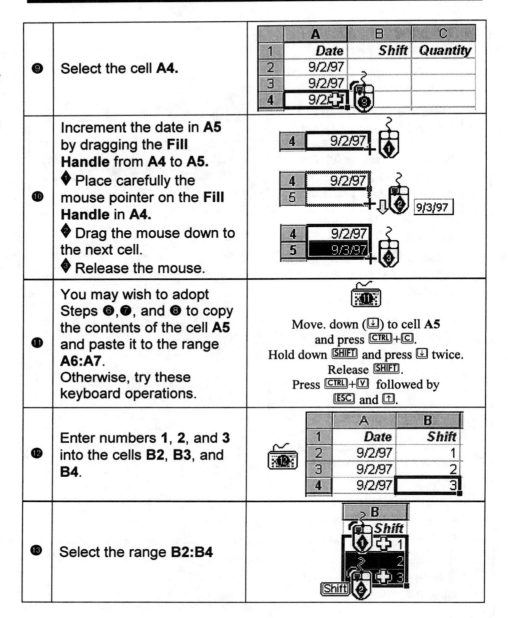

| ❾ | Select the cell **A4**. | |

| ❿ | Increment the date in **A5** by dragging the **Fill Handle** from **A4** to **A5**.
◆ Place carefully the mouse pointer on the **Fill Handle** in **A4**.
❷ Drag the mouse down to the next cell.
◆ Release the mouse. | |

| ⓫ | You may wish to adopt Steps ❻,❼, and ❽ to copy the contents of the cell **A5** and paste it to the range **A6:A7**.
Otherwise, try these keyboard operations. | Move. down (↓) to cell **A5** and press CTRL+C.
Hold down SHIFT and press ↓ twice. Release SHIFT.
Press CTRL+V followed by ESC and ↑. |

| ⓬ | Enter numbers **1**, **2**, and **3** into the cells **B2**, **B3**, and **B4**. | |

| ⓭ | Select the range **B2:B4** | |

Copy with Drag'n Drop:
The next step shows a popular drag'n drop operation applied to copying a range of cells. When you drag a Windows object while CTRL *key is held down, the object will be replicated at a new location as soon as you drop it there. Remember to first drop the object and then release* CTRL*. Note that in Windows 95 you can accomplish the same result by dragging an object with the right rather than left button pressed. When you drop the object, a menu pops up. You can then complete the operation by selecting* **Copy**, **Move** *or other commands.*

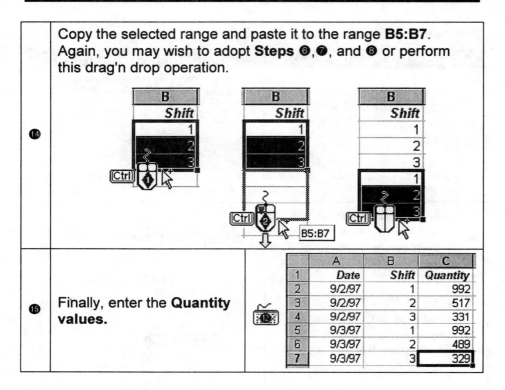

	Copy the selected range and paste it to the range **B5:B7**. Again, you may wish to adopt **Steps ⑥,⑦, and ⑧** or perform this drag'n drop operation.
⑭	

				A	B	C
⑮	Finally, enter the **Quantity values.**		1	Date	Shift	Quantity
			2	9/2/97	1	992
			3	9/2/97	2	517
			4	9/2/97	3	331
			5	9/3/97	1	992
			6	9/3/97	2	489
			7	9/3/97	3	329

Data entry tasks belong to the most time consuming and boring computer tasks. Since you have just completed such a task, it is a perfect time to save your work.

Section 3.4 Saving a Workbook

Computer systems are not perfect. They do fail from time to time, which should be at least one good reason not to wait with saving your work until all the work is done. It takes only a few seconds to save a workbook for the first time and a fraction of that time for subsequent saving operations.

The following procedure shows how to save this production quantity workbook as **ProdQty.xls**, onto your data diskette, in the folder **\Ch03**. Have your data diskette ready.

Procedure 3.4 Saving a Workbook File

	Task Description	Mouse / Keyboard
❶	On the **Formatting Toolbar**, click the **Save** button.	X Microsoft Excel - Book1 — File Edit View Insert — Arial Save
	Excel opens the **Save As** *dialog box.*	
❷	To open the list of drives and folders, click the **Drop-down** arrow, located at the **Save in:** box.	Save in: ☐ My Documents
❸	Make sure there is your data disk on the **Floppy** drive. Then, from the list, select the option **3½ Floppy**.	☐ My Documents — Desktop — My Computer — 3½ Floppy (A:) — (C:) — My Documer
❹	Switch to the **Ch03** folder.	Save As — Save in: 3½ Floppy (A:) — Ch02 C — Ch03 — Ch0 — Ch05
❺	In the **File name:** box, double-click the name **Book1**.	File name: Book1.xls
❻	Type the new name, **ProdQty**, over the selected one.	File name: ProdQty.xls
❼	Click the **Save** button.	Save

Note:
We recommend that you use a copy of the original data diskette. Procedure 2.2 shows how to duplicate a diskette.

Note:
In Windows text boxes or word processors, a common way to select (highlight) a single word is to double-click it.

From now on, to save the **ProdQty.xls** workbook again, you only need to do **Step ❶**.

Section 3.5 Using Named References in Excel

Since column ranges are used for different variables in a data set, it would be convenient to refer to each variable by a name rather than by the range reference. For example, in order to calculate the sum of all *Quantity* values, you would use the **Sum**( ) function with the reference **C2:C7**:

=Sum(C2:C7).

With the range **C2:C7** named as **Quantity**, the function would be:

=Sum(Quantity),

which is definitely more informative and also self-documenting.

Since the first (top) row of the data set usually contains names or some other identifiers of the variables (columns), it makes perfect sense to use them as the reference names. For example, the data set shown in Figure 3.2 contains three variables labeled as *Date*, *Shift*, and *Quantity*. The following procedure shows how to use these labels as names of their corresponding ranges **A2:A7**, **B2:B7**, and **C2:C7**.

Procedure 3.5 Creating Column-range Names

	Task Description	Mouse / Keyboard
❶	Select the range containing both the names and values of the variables (here **A1:C7**). ◆ Click **A1** and ◆ shift-click **C7**.	
❷	Invoke the ◆ **Insert** ◆ **Name** ◆ **Create** command.	
	*Excel opens the **Create Names** dialog box.*	

❸	◆ Check off the **Left column** box and ◆ click the OK button.

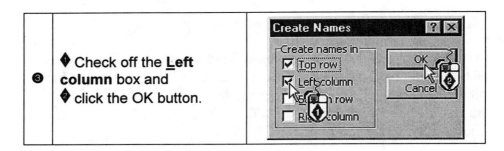

To learn how the names can be used, you will calculate the total number of shifts and the sum of all quantities using the **Count( )** function and the **Sum( )** function, respectively.

Procedure 3.6   Using the Count( ) and Sum( ) Functions with Named References

	Task Description	Keyboard
❶	Type **Total** in the cell **A8** and press ⮐ to move to the cell **B8**.		A	B	C | 1	Date	Shift	Quantity | 2	9/2/97	1	992 | 3	9/2/97	2	517 | 4	9/2/97	3	331 | 5	9/3/97	1	992 | 6	9/3/97	2	489 | 7	9/3/97	3	329 | 8	Total		

Note:
The function you enter in a cell is shown on the **Formula** *bar, while its value is returned in the cell.*

❷	Type the **=Count(Shift)** function and press CTRL+ENTER.

B8 ▼ = =COUNT(Shift)

	A	B	C	D
1	Date	Shift	Quantity	
2	9/2/97	1	992	
3	9/2/97	2	517	
4	9/2/97	3	331	
5	9/3/97	1	992	
6	9/3/97	2	489	
7	9/3/97	3	329	
8	Total	6		

❸

Press ↵ to move to the cell **C8**, type the **=Sum(Quantity)** function, and press CTRL + ENTER .

C8	▼	=	=SUM(Quantity)

	A	B	C	D
1	*Date*	*Shift*	*Quantity*	
2	9/2/97	1	992	
3	9/2/97	2	517	
4	9/2/97	3	331	
5	9/3/97	1	992	
6	9/3/97	2	489	
7	9/3/97	3	329	
8	Total		6	3650

❹

To select the range **A8:C8**, hold down SHIFT and press ← twice.

8	Total		6	3650

❺

Press CTRL + B to **Bold** the selection. Then press CTRL + HOME to arrive at **A1**.

A1	▼	=	Date

	A	B	C
1	*Date*	*Shift*	*Quantity*
2	9/2/97	1	992
3	9/2/97	2	517
4	9/2/97	3	331
5	9/3/97	1	992
6	9/3/97	2	489
7	9/3/97	3	329
8	**Total**		**6**	**3650**

Note:
*It is a good time to save your work. Just click the **Save** button:*

Section 3.6 Printing Your Work

Printing a worksheet in Excel can be as easy as clicking the **Print** button. When you click this button, Excel will start printing the sheet you are in, using all current settings. To customize your printout, you need to execute the **Page Setup** command or the **Print** command from the **File** menu.

Note:
The shortcut for the **File | Print** *command is this Toolbar button:*

Suppose that you want to print the **ProdQty.xls** worksheet (**Sheet1**) featuring:

- ▪ *Gridlines,*
- ▪ *Worksheet Borders* (column headings and row numbers),
- ▪ *Page Header* with the title Production Quantities and current Date,
- ▪ *Page Footer* with the file name **ProdQty.xls**, and
- ▪ *Page Centered* horizontally.

After you have completed you job, the printout should look approximately like this one:

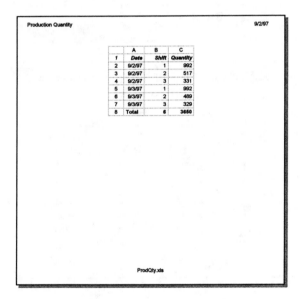

Here is how to do it:

Procedure 3.7 Printing a Sheet

Task Description	Mouse/Keyboard
❶ On the **Formatting Toolbar**, click the **Print Preview** button.	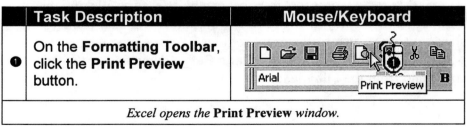
*Excel opens the **Print Preview** window.*	

❷	Click the **Setup...** button.	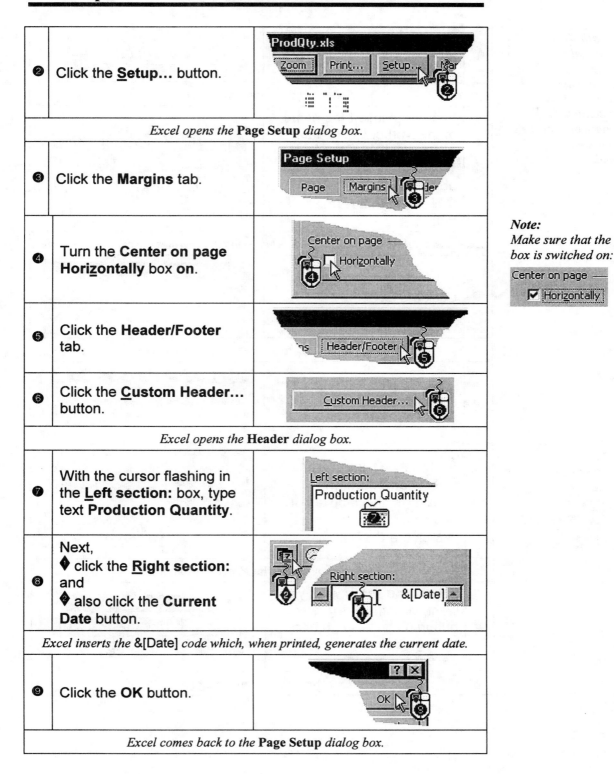
	Excel opens the **Page Setup** *dialog box.*	
❸	Click the **Margins** tab.	
❹	Turn the **Center on page Horizontally** box **on**.	
❺	Click the **Header/Footer** tab.	
❻	Click the **Custom Header...** button.	
	Excel opens the **Header** *dialog box.*	
❼	With the cursor flashing in the **Left section:** box, type text **Production Quantity**.	
❽	Next, ◆ click the **Right section:** and ◆ also click the **Current Date** button.	
	Excel inserts the &[Date] *code which, when printed, generates the current date.*	
❾	Click the **OK** button.	
	Excel comes back to the **Page Setup** *dialog box.*	

Note:
Make sure that the box is switched on:

Center on page
☑ Horizontally

⑩	Now, click the **Custom Footer** button.	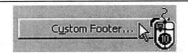
	Excel opens the **Footer** *dialog box.*	
⑪	In order to insert the name of the workbook into the center section of the footer, ◆ click the **Center section:** box, and ◆ click the **File Name** button.	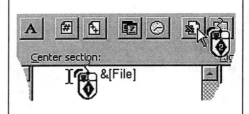
	Excel inserts the &[File] *code which, when printed, generates the current file name.*	
⑫	Click the **OK**.	
	Excel comes back to the **Page Setup** *dialog box.*	
⑬	Click the **Sheet** tab.	
⑭	◆ Turn the **Gridlines** and ◆ **Row and column headings** check boxes **on** and ◆ click the **OK** button.	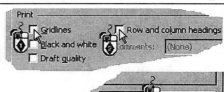
	Excel shows the **Print Preview** *window.*	
⑮	If what you see is what you want, click the **Print...** button. Otherwise go back to Page Setup and make necessary changes.	
	Excel opens the **Print** *dialog box.*	

⑯	If all settings are acceptable, click the **OK** button.	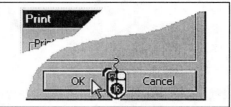

Note:

*If you want to print a specific range, rather than the entire sheet, you must select this range prior to invoking one of the **Print** related commands. Then, in the **Print** dialog box, you would also have to choose the **Selection** option button.*

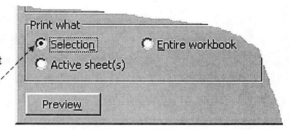

It is again a good time to save your workbook. Note that all print settings will also be saved.

Section 3.7 Exiting Excel

Closing an Excel window requires the same operations as closing most of the other windows. You can double-click the window's **Control Menu** box, click the **Close** button, or execute the **File | Exit** command. If any of the workbooks you worked with has not been saved, Excel will open its closing message box as shown below.

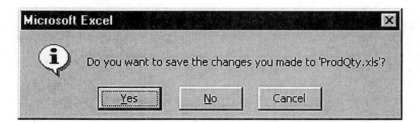

Take any action you deem appropriate. Note that selecting the **Cancel** button will abandon the exit operation.

Chapter 4   "Complaint Department"

Displaying Qualitative Data

Section 4.1 Overview

Statistical Objectives: After reading this chapter and doing the exercises you will:
- Know the function of a column chart, clustered bar chart and stacked column chart.
- Know the function of a pie chart.
- Be able to decide when a pie chart or a column chart is more appropriate.
- Know when a clustered column chart is better than multiple charts.
- Know how to examine column charts for trends and patterns.
- Know how to draw conclusions from the information obtained from graphical displays of qualitative data.

Section 4.2 Problem Statement

Most large companies that manufacture consumer goods receive feedback from the customers. A large manufacturer of paper goods keeps track of consumer complaints for its facial tissue product line. The company receives these complaints through a toll free number that appears on the product. The data taken consists of a transcription of the actual complaint language and a classification of the complaint into a specific category. When the company gets a complaint they generally ask for additional information about the specific package. They then use the information to relate the complaints to manufacturing data so that in the future similar problems can be prevented.

Table 4.1 lists the categories and sub categories used by the company to classify the complaints:

Table 4.1 Complaint Categories

Category	Sub category
Dispensing	Sheets Tear on Removal
	Reach In/Fallback
Foreign Materials	Lint/Dust
	Other
Odor	
Miscounts	
Packaging	Defective
	Misleading
	Damaged
	Other

For example, if a consumer calls the company and says "...the box of tissues I bought was not full...," the customer service representative will classify this as a miscount. Sometimes complaints can be tricky to classify. If a customer calls, for example, after a product change and says "... I eat tissues and I prefer the taste of the old <product>..." into what category should they place the complaint?

Section 4.3 Characteristics of the Data Set

FILENAME: Ch04Dat.xls   An Excel Workbook
SIZE: COLUMNS   5
                   ROWS   361

The first 7 lines of the data file along with the variable names are shown in Figure 4.1.

	A	B	C	D	E
1	*Month*	*Year*	*Category*	*Subcategory*	*Number*
2	JANUARY	1989	DISPENSING	FALLBACK	80
3	JANUARY	1989	DISPENSING	SHEETS_TEAR	118
4	JANUARY	1989	FOREIGN MATL.	LINT/DUST	53
5	JANUARY	1989	FOREIGN MATL.	OTHER	1
6	JANUARY	1989	MISCOUNTS	MISCOUNTS	37
7	JANUARY	1989	ODOR	ODOR	5

Ch04Dat.xls

Data / Sheet2 / Sheet3 / Sheet4 / Sl

Figure 4.1 The Complaint Data File

Start Excel and open the **Ch04Dat.xls** data file. Refer to Procedure 3.1 (page 33) for help on how to open a workbook. It is recommended that you use a copy of your original diskette. From Procedure 2.2 (page 22), you can learn how to duplicate the diskette.

In Excel, click the Open button. Then select the drive, folder, and file name.

Section 4.4 Overview of Graphs in Excel

Excel is equipped with a user-friendly *assistant* for creating graphs called **Chart Wizard**. Excel can create graphs on a separate sheet or embed it onto a data sheet (worksheet). In creating graphs, mouse operations are more convenient. Nevertheless, Excel is capable of creating graphs using the keyboard exclusively

Chart Wizard *belongs to the* **Standard Toolbar**.

The following steps summarize a typical way to create an Excel chart:

■ Select (highlight) data to be graphed.
■ Invoke the **Chart Wizard**.
■ Follow the steps provided by the **Chart Wizard**.
■ Edit, annotate, size and position the chart on the worksheet.

After the graph has been created, it can be edited, annotated, sized, copied, moved, and otherwise customized in many possible ways. It can be printed and saved alone or with data.

Section 4.5 How to Create Column Charts in Excel

Excel distinguishes between *bar* and *column* charts. Charts that show vertical bars are referred to as *column* charts. Those displaying horizontal bars are called *bar* charts. This section focuses on the basics of creating column charts using Excel. Examples of some of the more general features of graphs such as titles and formatting are also covered.

A Bar Chart

A Column Chart

Your task is to create a chart approximately the same as the one shown in Figure 4.2. This chart has the following attributes:

X-axis:	The 10 *Subcategory* classes
Y-axis, Series 1:	The 10 *Number* values for January 1989
Chart Type:	Default (Column)
Chart Format:	Default (Clustered Column)
Titles:	
Chart:	**Customer Complaint** **January 1989**
X-axis:	**Complaint Subcategories**
Y-axis:	**Frequency**
Legend:	None
Annotations:	Text boxes and arrows

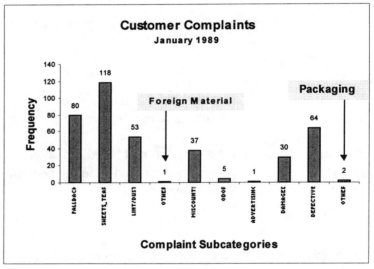

Figure 4.2 Column Chart of Customer Complaints for January 1989

Note that the subcategory OTHER appears twice among complaint subcategories. The first one relates to the category FOREIGN MATERIALS, the second one to PACKAGING. This is why the chart needs to be annotated to provide reliable information. We will use two drawing objects, **Text** box and **Arrow** for that purpose.

Section 4.5.1 Creating a Column Chart

When creating a column chart, Excel expects to find the classifications for the data organized by columns or rows. We will begin by creating a column chart to display customer complaints for January 1989. The data is already in the correct format for Excel to produce such a graph with a few simple operations.

Procedure 4.1 Generating a Column Chart

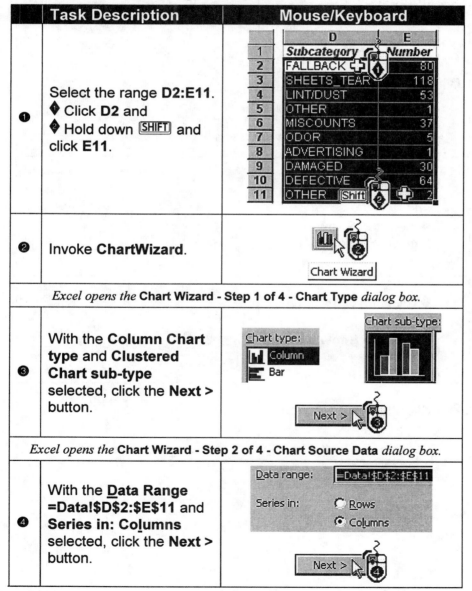

	Task Description	Mouse/Keyboard
❶	Select the range **D2:E11**. ◆ Click **D2** and ◆ Hold down SHIFT and click **E11**.	
❷	Invoke **ChartWizard**.	
	Excel opens the **Chart Wizard - Step 1 of 4 - Chart Type** *dialog box.*	
❸	With the **Column Chart type** and **Clustered Chart sub-type** selected, click the **Next >** button.	
	Excel opens the **Chart Wizard - Step 2 of 4 - Chart Source Data** *dialog box.*	
❹	With the **Data Range =Data!$D$2:$E$11** and **Series in: Columns** selected, click the **Next >** button.	

Trouble!
If you do not see the **Chart Wizard** *button, maximize the Excel program window. If the* **Standard Toolbar** *is not displayed, run the* **View | Toolbars | Standard** *command.*

Excel opens the **Chart Wizard - Step 3 of 4 - Chart Options** *dialog box.*

Note:

Carefully observe the sample chart box to see how the options you select affect your chart.

⑤ In the **Titles** section, click the **Chart title** box and type **Customer Complaints January 1989**. Next, press TAB and type **Complaint Subcategories**. Finally, press TAB and type **Frequency**.

Make sure the **Major gridlines** *check box is empty:*

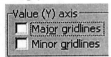

⑥ ❶ Click the **Gridlines** tab and
❷ clear the **Major gridlines** check box.

Make sure the **Show legend** *check box is empty:*

❏ Show legend

⑦ ❶ Click the **Legend** tab and
❷ clear the **Show legend** check box.

Make sure the **Show value** *option button is turned on:*

⑧ ❶ Click the **Data Labels** tab and
❷ click the **Show value** check box.

⑨ Click the **Next >** button.

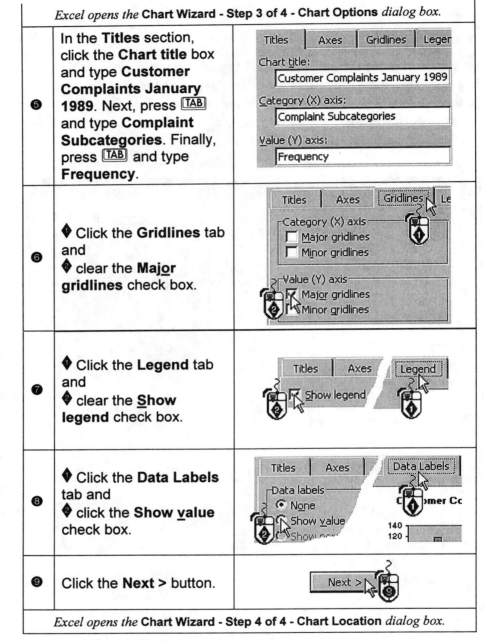

Excel opens the **Chart Wizard - Step 4 of 4 - Chart Location** *dialog box.*

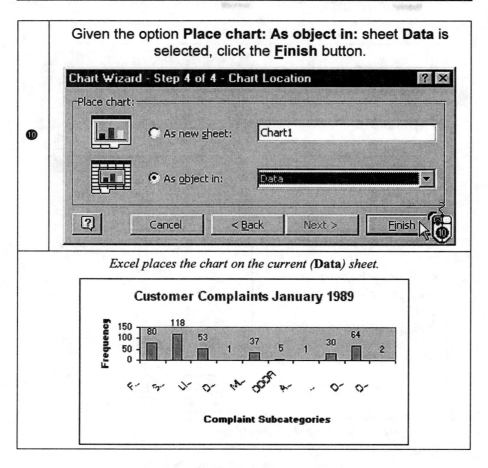

Given the option **Place chart: As object in:** sheet **Data** is selected, click the **Finish** button.

*Excel places the chart on the current (**Data**) sheet.*

The chart needs a little more work. We are going to change the format of the **Category (X) Axis** labels, eliminate the plot area background, enlarge the chart, modify **Chart Title**, change some fonts, and annotate the two **OTHER** categories.

Procedure 4.2 Formatting the Category (X) Axis

	Task Description	Mouse/Keyboard
❶	Click the chart anywhere to select it.	
❷	Double-click the **Category (X) Axis** area.	

Note:

A selected chart or drawing object shows little black boxes (handles) along its perimeter.

The handles can be dragged to resize the chart.

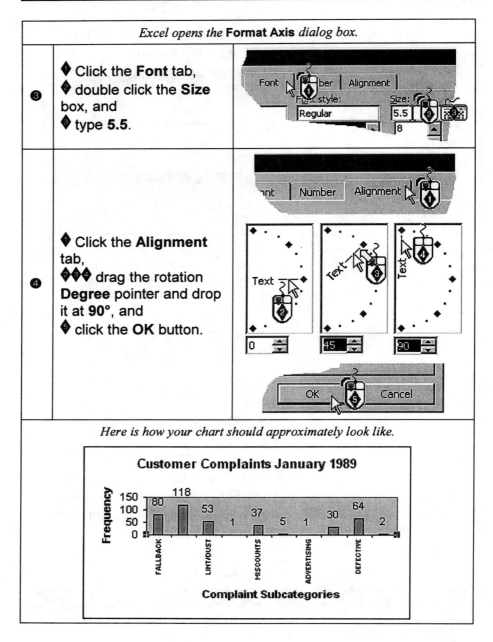

Excel opens the **Format Axis** *dialog box.*

③
◆ Click the **Font** tab,
◆ double click the **Size** box, and
◆ type **5.5**.

④
◆ Click the **Alignment** tab,
◆◆◆ drag the rotation **Degree** pointer and drop it at **90°**, and
◆ click the **OK** button.

Here is how your chart should approximately look like.

Customer Complaints January 1989

The following procedure shows how to clear the **Plot** area.

Procedure 4.3 Clearing the Plot Area Background

	Task Description	Mouse/Keyboard
❶	Double click the **Plot** area.	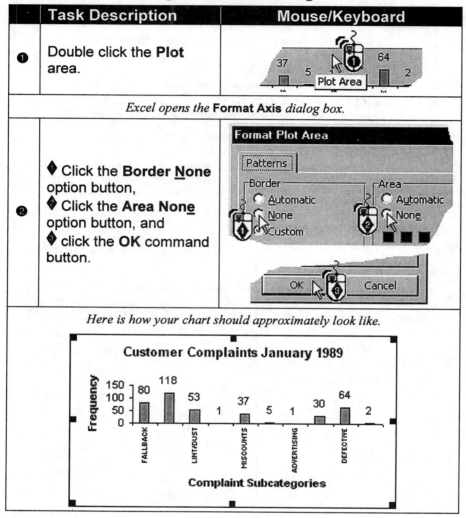
	Excel opens the **Format Axis** *dialog box.*	
❷	◆ Click the **Border None** option button, ◆ Click the **Area None** option button, and ◆ click the **OK** command button.	

Make sure the **Border None** *and* **Area None** *option buttons are turned on:*

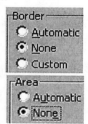

Here is how your chart should approximately look like.

The chart is too small to show clearly all its components. In order to enlarge the chart, drag its lower-right-corner **Handle** diagonally right and down. ☞

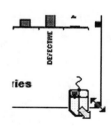

Your next formatting task is to modify the chart title so it will look like this one:

Customer Complaints
January 1989

Procedure 4.4 Reformatting the Chart Title

	Task Description	Mouse/Keyboard
❶	Click the **Chart Title** box, in order to select it. Customer Complaints January 1989 — Chart Title	
❷	Place the **I-beam** mouse pointer right in front of word **January** and click the mouse again.	...ai[ ]January 1989
❸	Press ENTER to push the right-hand side text down to the next line.	Enter January 1989
❹	Hold down SHIFT and press END, in order to select the second line.	Customer Complaints Shift End January 1989
❺	On the **Formatting Toolbar**, click **the Font Size Drop-down** arrow.	Arial ▼ 15 ▼ I — Chart Title ▼ — Font Size
❻	Select **Size 10**.	Arial ▼ 15 ▼ — Chart Title ▼ — 8 9 A B 10 1 Month Yea 11 2 JANUARY 198! 12 3 JANUARY 198! 14 4 IANI 16
❼	Click anywhere outside the chart, in order to un-select it.	A B C D 1 Month Year Category Subcategory 2 JANUARY 1989 DISPENSING FALLBACK 3 JA 4 JA 5 JA Customer Complaints 6 JA January 1989 7 JA 8 JA 140 118 9 JA 120

Generally, in order to change the format of a chart object, you usually point to the object and either double-click (🖱) or right-click (🖱) the mouse. A double-click operation brings you to a dialog box that deals with the most popular attributes of the object (see step ❷ of Procedure 4.2 or step ❶ of Procedure 4.3). On the other hand, a right-click action (👆) first opens a shortcut menu, related to the object. The following procedure shows the latter at work.

Procedure 4.5 Changing the Font Size of the Data Labels

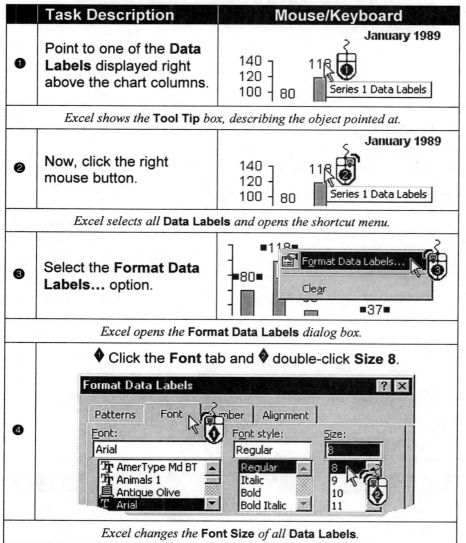

	Task Description	Mouse/Keyboard
❶	Point to one of the **Data Labels** displayed right above the chart columns.	January 1989 ... Series 1 Data Labels
	Excel shows the **Tool Tip** *box, describing the object pointed at.*	
❷	Now, click the right mouse button.	January 1989 ... Series 1 Data Labels
	Excel selects all **Data Labels** *and opens the shortcut menu.*	
❸	Select the **Format Data Labels...** option.	Format Data Labels... Clear
	Excel opens the **Format Data Labels** *dialog box.*	
❹	◆ Click the **Font** tab and ◆ double-click **Size 8**.	Format Data Labels ...
	Excel changes the **Font Size** *of all* **Data Labels**.	

Note:

Given you want to format the **Data Labels**, *instead of the three steps, you can do just one double click:*

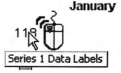

Note:

Frequently, when you double-click an option in a dialog box, such an action will both select the option and close the dialog box.

Font attributes of the text objects can also be changed using the **Formatting Toolbar**. The following procedure shows how to change the font **Size** of the **Value (Y) Axis** labels.

Procedure 4.6 Changing the Font Size of the Value Axis Labels

Task Description	Mouse/Keyboard
❶ Click any of the **Value Axis** labels.	
❷ On the **Formatting Toolbar**, click **the Font Size Drop-down** arrow.	
❸ Select **Size 8**.	

Your last chart-editing task is to annotate the values of the OTHER subcategories, using two drawing objects: **Text Box** and **Arrow**. Check first if the **Drawing Toolbar** is open:

It is usually docked (located) at the bottom of the Excel window. The following procedure shows how to open the **Toolbar**.

Note:

*In Excel, a toolbar can be moved to another location by dragging the **Move Handle** at the left side of a docked toolbar, or by dragging the **Title Bar** on a floating toolbar.*

Procedure 4.7 Opening a Toolbar

Task Description	Mouse/Keyboard
❶ If the **Drawing Toolbar** is not open, execute the ◆ **View** ❷ **Toolbars** ◆ **Drawing** command.	
*Make sure the **Drawing** option gets turned on:*	

Procedure 4.8 Embedding a Text Box

	Task Description	Mouse/Keyboard
❶	On the **Drawing Toolbar**, click the **Text Box** button.	
❷	Place the **Cross Pointer** approximately as shown at the right.	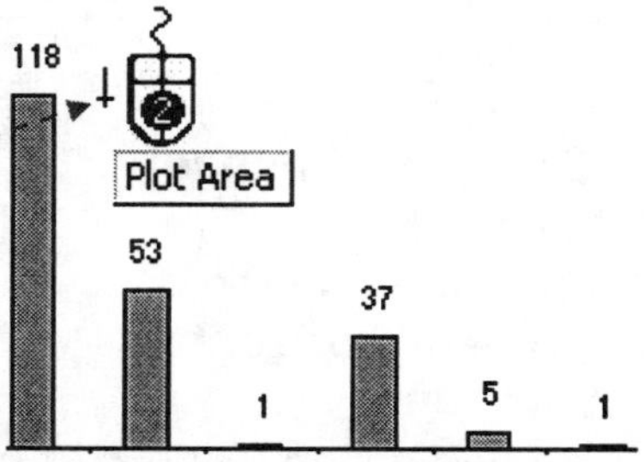
❸	Drag the mouse diagonally to the right and down.	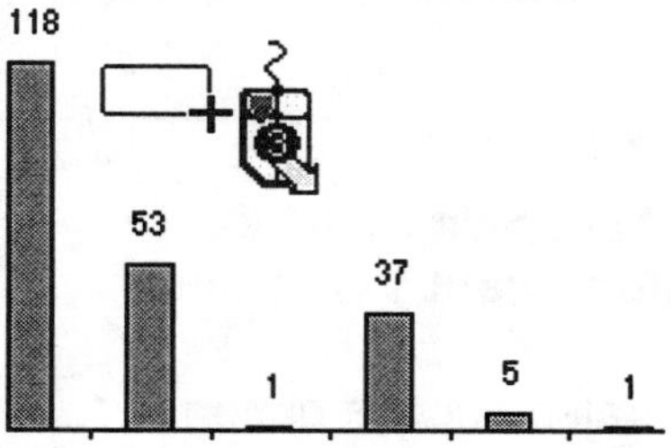
❹	Drop the lower-right corner of the box as shown at the right.	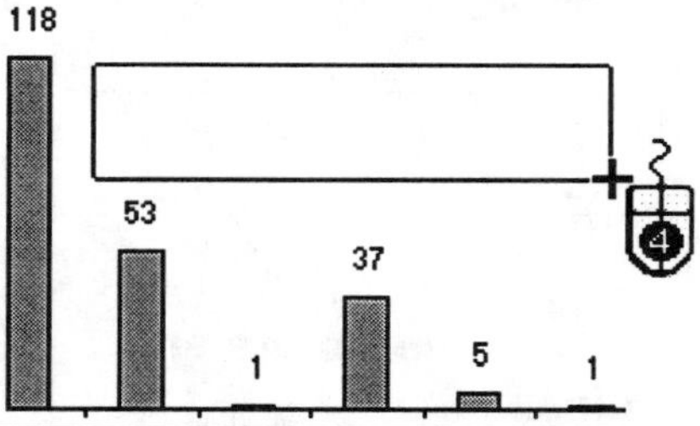
❺	Right away, type text **Foreign Material** (<u>do not</u> press ENTER).	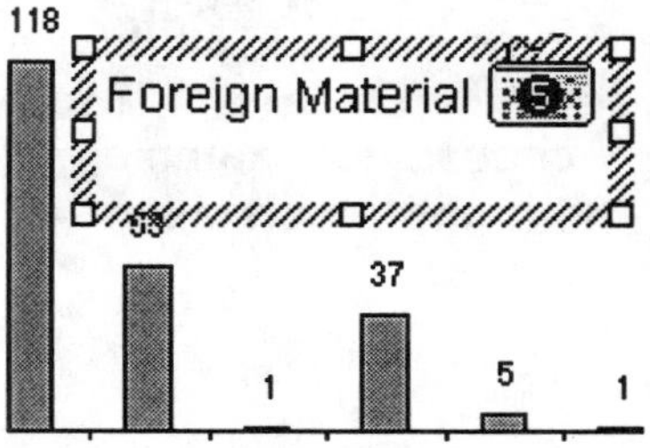
❻	Place the mouse pointer on the **Text Box** border and double-click it.	
	*Excel opens the **Format Text Box** dialog box.*	

Note:

The location and size of the **Text Box** *can be changed later.*

Note:

Do not worry if there is not enough room for the test. The **Text Box** *can be enlarged later by dragging one of its* **Sizing Handles**.

⑦	◆ Click the **Font** tab, ◆ choose **Font Style Bold**, and ◆ pick font **Size** 8.	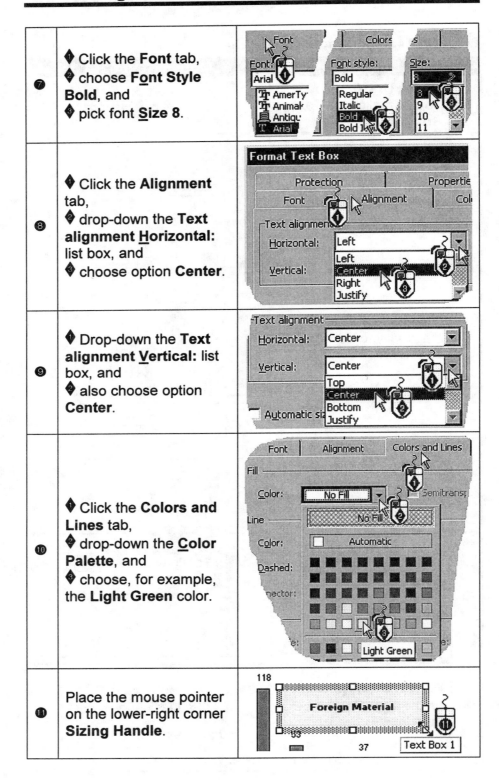
⑧	◆ Click the **Alignment** tab, ◆ drop-down the **Text alignment Horizontal:** list box, and ◆ choose option **Center**.	
⑨	◆ Drop-down the **Text alignment Vertical:** list box, and ◆ also choose option **Center**.	
⑩	◆ Click the **Colors and Lines** tab, ◆ drop-down the **Color Palette**, and ◆ choose, for example, the **Light Green** color.	
⑪	Place the mouse pointer on the lower-right corner **Sizing Handle**.	

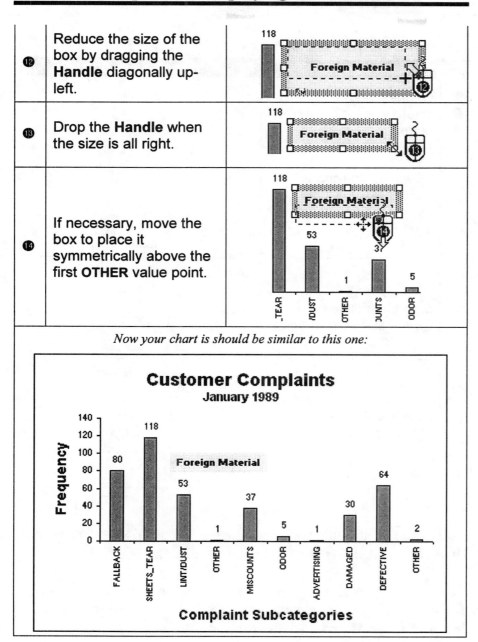

⑫	Reduce the size of the box by dragging the **Handle** diagonally up-left.	
⑬	Drop the **Handle** when the size is all right.	
⑭	If necessary, move the box to place it symmetrically above the first **OTHER** value point.	

Now your chart is should be similar to this one:

Note:

*In order to move a **Text Box** or other framed object, you must drag the frame itself. Make sure you do not touch the **Sizing Handles**.*

Your next job is to draw an arrow going from the **Text Box** to the **Data Label** of the **OTHER** subcategory.

Procedure 4.9 Drawing an Arrow

	Task Description	Mouse/Keyboard
❶	On the **Drawing Toolbar**, click the **Arrow** button.	
❷	Place the drawing pointer in the middle of the lower edge of the **Text Box**.	
❸	Hold down SHIFT and drag the mouse down toward **Data Label 1**.	
❹	Drop the **Arrow Head** right above the label.	

Note:

In order to draw a straight line (horizontal, vertical, or diagonal), press and hold down SHIFT *while you are drawing the line.*

Note:

To copy a **Text Box***, drag its border and drop it at the target location while* CTRL *is being held down. Or you may wish to try the right-drag mouse action.*

Exercise 1. In a similar way, annotate the second **OTHER** data point. This time the **Text Box** should contain word **Packaging**.

Hint: Rather than drawing, typing, and formatting a brand new **Text Box**, copy and paste the existing one. You will only need to replace the text **Foreign Material** with **Packaging**.

When done, your chart should look like the one shown in Figure 4.2.

Exercise 2. What does the bar graph tell you about the distribution of customer complaints? Which category is the highest? Which one is the lowest?

Section 4.5.2 Printing a Chart in Excel

In order to print a chart created on a separate (chart) sheet, move to the sheet and click the **Print** icon or press CTRL+P, ENTER. Excel will print the chart according to its default settings.

For a customized printout, you need to run the **Page Setup** command. You can invoke this command directly from the **File** menu (ALT+F, U), from the **Print** dialog box, or from the **Print Preview** window (preferred). It is also possible to visually manipulate some page settings while in the **Preview** window.

Whether you print a sheet or selected range, it is always a good practice to start the printing process from the **Print Preview** window!

Procedure 3.7 (page 44) provides an exhaustive example of printing one data sheet. Most of the commands, demonstrated there, also apply to printing Excel charts.

Printing a chart with embedded drawing objects (text boxes, lines, rectangles, arrows, etc.) may be a little tricky chore. For example, if you draw some objects on a chart embedded in a data

Note:

*Unless you want to print an entire sheet, **do not** use the **Print** button:*

Print

Print Preview

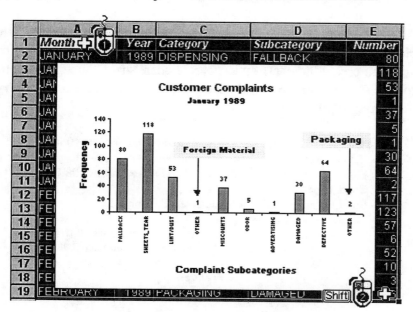

Figure 4.3 Selecting a Range of Cells and a Chart

sheet (as you did with the above column chart) and then print the chart

Note:

It is a good time to save your workbook:

alone, the drawing objects will not be printed. In order for the object to be printed, you would have to select worksheet cells along with the chart and drawing objects. Figure 4.3 shows an example of such a selection. On the other hand, if your chart resides on a **Chart** sheet, then all drawing objects will be printed along with the chart.

Section 4.6 Clustered and Stacked Column Charts

Comparing data for several different scenarios is something that management needs to do in order to make sound business decisions. If, for example, the company wanted to know whether complaints for the last month were comparable to previous months, they could look at bar charts for the months of interest together to get this information. In this section you will learn how to view your data using **Clustered** and **Stacked** column charts.

Section 4.6.1 Clustered Column Charts

When you have similar data for different situations, you can cluster the bars or columns together for the same categories, so that the data from all of the different situations appear on the same graph.

Again, you will use the **Chart Wizard**, in order to generate a clustered column chart for the first two months of 1989. This time you will create the chart on a separate chart sheet. The specifications for the **Customer Complaint Distribution** cluster chart for **January** and **February, 1989**, are as follows:

X-axis:	Variable *Subcategory* (the 10 labels)
Y-axis, Series 1:	Variable *Number* (in rows for January 1989)
Y-axis, Series 2:	Variable *Number* (in rows for February 1989)
Chart Type:	Column
Chart Format:	Clustered Column
Titles:	
Chart:	**Customer Complaint Distribution**
	January and February 1989
X-axis:	**Complaint Subcategories**
Y-axis:	**Frequency**
Legend:	January, February

Figure 4.4 shows the final outcome.

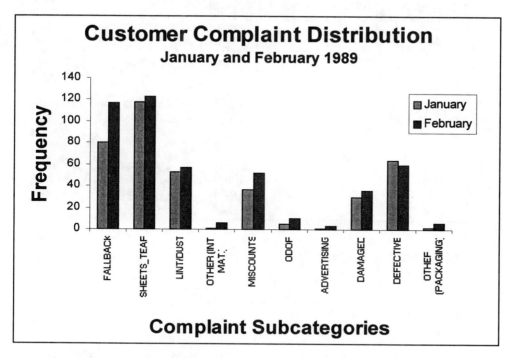

Figure 4.4 Clustered Column Chart for Customer Complaints

The input data for the chart resides on the **Data** sheet, in cell ranges **D1:E11** and **E12:E21**. It will be more convenient to create and analyze the chart with the data organized as a rectangular block. Figure 4.5 shows the input data for the chart, located on the **ClusteredChart** sheet. The steps that follow show how to create it.

	A	B	C	D	E
1				Number	
2			Subcategory	January	February
3			FALLBACK	80	117
4			SHEETS_TEAR	118	123
5			LINT/DUST	53	57
6			OTHER	1	6
7			MISCOUNTS	37	52
8			ODOR	5	10
9			ADVERTISING	1	3
10			DAMAGED	30	36
11			DEFECTIVE	64	60
12			OTHER	2	6

|◀ ◀ ▶ ▶| Data \ **ClusteredChart** / Sheet3 / Sheet4 / Sheet5 /

Figure 4.5 Customer Complaints for January and February

Procedure 4.10 Preparing Data for the Clustered Chart

	Task Description	Mouse/Keyboard
❶	On the **Data** sheet, select range **D1:E11**.	
❷	Click the **Copy** button.	
	Excel copies the selected range to **Clipboard**.	
❸	Switch to **Sheet2**.	
❹	Select cell **C2**.	
❺	Click the **Paste** button.	
	Excel pastes the contents of **Clipboard** *onto the worksheet, starting at cell* **C2**.	

⑥	Rename **Sheet2** to **ClusteredChart**. ◆ Double-click the **Sheet2** tab and ❷ enter text **ClusteredChart**.	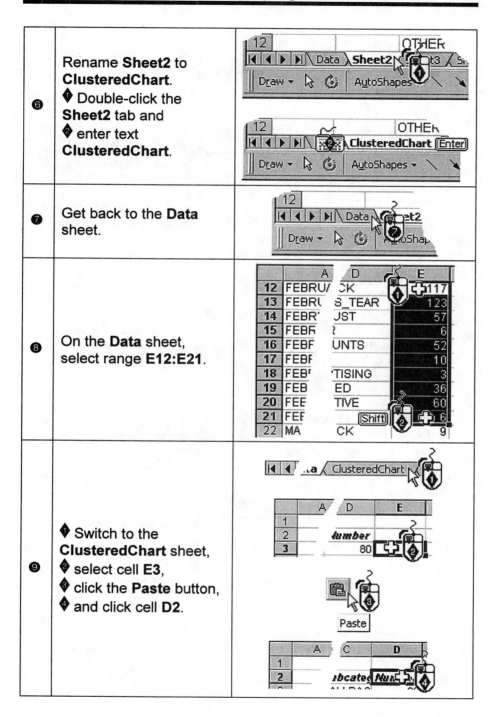
⑦	Get back to the **Data** sheet.	
⑧	On the **Data** sheet, select range **E12:E21**.	
⑨	◆ Switch to the **ClusteredChart** sheet, ❷ select cell **E3**, ❸ click the **Paste** button, ❹ and click cell **D2**.	

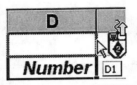

⑩ Move the contents of cell **D2** to **D1**.

◆ Place the mouse pointer on the **Selection** border of cell **D2** then press and hold down the **left** mouse button.

❷ Drag the selection up.

❸ Drop it onto **D1**.

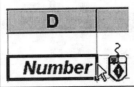 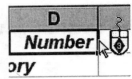

Note: *When you drag a cell or range of cells with the **right** mouse button and then drop it at the target location, Excel opens a shortcut menu as shown at the right. You can then decide how to complete this drag'n drop operation.*

Move Here
Copy Here
Copy Here as Values Only
Copy Here as Formats Only
Link Here
Create Hyperlink Here

Shift Down and Copy
Shift Right and Copy
Shift Down and Move
Shift Right and Move

Cancel

⑪ In order to automatically adjust the width of column **C**, place the mouse pointer between the **Headings** of columns **C** and **D** and double-click the **left** mouse button.

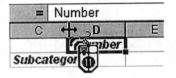

⑫ ◆ Type word **January** into cell **D2**,

❷ click the **Enter** button, and

❸ change the **Font Style** to **Bold**.

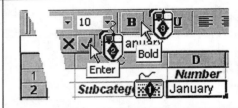

Note:

*By dragging the **Fill Handle**, you can produce typical sequential list (dates, months, days, numeric series, etc.)*

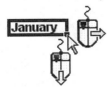

⑬ Drag the **Fill Handle** of **D2** and drop it onto cell **E2**.

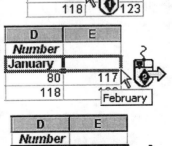

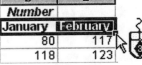

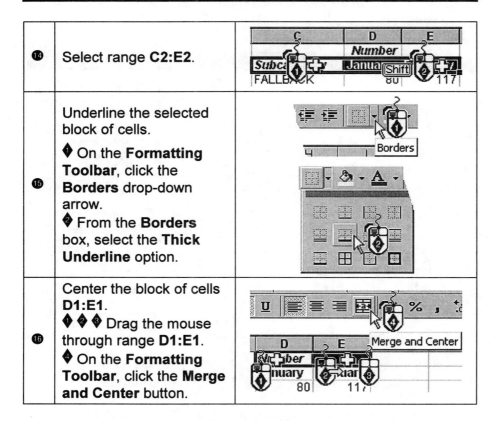

⑭	Select range **C2:E2**.	
⑮	Underline the selected block of cells. ◆ On the **Formatting Toolbar**, click the **Borders** drop-down arrow. ❷ From the **Borders** box, select the **Thick Underline** option.	
⑯	Center the block of cells **D1:E1**. ◆ ◆ ◆ Drag the mouse through range **D1:E1**. ◆ On the **Formatting Toolbar**, click the **Merge and Center** button.	

Note:

It is a good time to save your workbook:

You may wish to compare your chart with the one shown in Figure 4.5. Note that the format features do not have to be identical and you can experiment with them on your own. However, make sure that the contents and structure of your **ClusteredChart** sheet are the same as in Figure 4.5.

Notice that also in this case the two **OTHER** subcategories create a potentially unclear situation. In the previous case, you solved this problem by properly annotating the chart. This time, however, you will change the sub categories right in their cells (**C6** and **C12**).

Procedure 4.11 Editing a Cell

	Task Description	Mouse/Keyboard
❶	Double-click cell **C6** in order to select and enable editing of the cell.	
❷	Press **Space Bar**, type **(INT. MAT.)**, and press ⎡ENTER⎤.	

Note:
Check step ⓫ *of Procedure* 4.10 *to see how to adjust the column width.*

In a similar way change the contents of cell **C12** to **OTHER** (**PACKAGING**). If necessary, adjust the width of column **C**.

You are now ready to create the chart. The steps are pretty similar to those described in Procedure 4.1. This is why some of the steps in following procedure refer frequently to Procedure 4.1.

Procedure 4.12 Generating a Clustered Column Chart

Note:
The **Value** *data labels (***January** *and* **February***), included in the chart data range, will be used by the* **Chart Wizard** *to create the* **Legend Box***.*

	Task Description	Mouse/Keyboard
❶	Select the chart data range (**C2:E12**).	
❷	Invoke the **Chart Wizard**.	

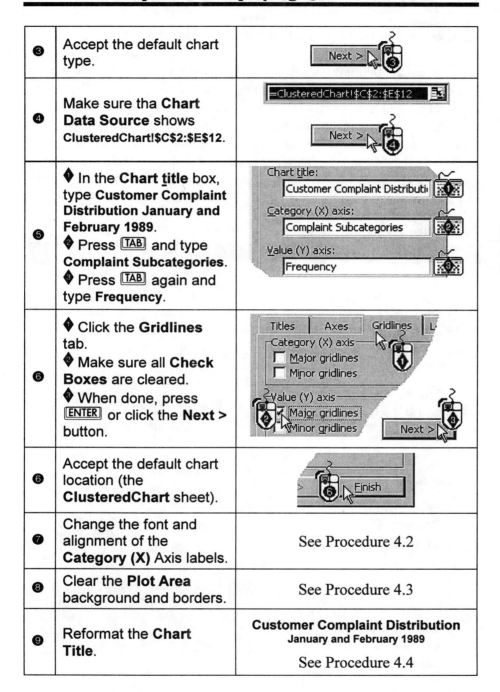

③	Accept the default chart type.	
④	Make sure tha **Chart Data Source** shows **ClusteredChart!$C$2:$E$12**.	
⑤	◆ In the **Chart title** box, type **Customer Complaint Distribution January and February 1989**. ◆ Press TAB and type **Complaint Subcategories**. ◆ Press TAB again and type **Frequency**.	
⑥	◆ Click the **Gridlines** tab. ◆ Make sure all **Check Boxes** are cleared. ◆ When done, press ENTER or click the **Next >** button.	
⑥	Accept the default chart location (the **ClusteredChart** sheet).	
❼	Change the font and alignment of the **Category (X)** Axis labels.	See Procedure 4.2
❽	Clear the **Plot Area** background and borders.	See Procedure 4.3
❾	Reformat the **Chart Title**.	**Customer Complaint Distribution** **January and February 1989** See Procedure 4.4

Note:
DO NOT *press* ENTER
until all chart titles
are defined.

Exercise 3. Create clustered column charts of the customer complaints for January and February of 1990.

Section 4.6.2 Creating Stacked Column Charts

Rather than having the columns of the chart clustered so that you can do direct comparisons among months or look at distributions by months, you may be interested in another way of looking at the information you have. A **Stacked Column** chart can be used to see the way each month contributes to the overall total number of complaints in a particular category. To create a **Stacked Column** chart with the same data you just used for the **Clustered Column** chart, you only need to make one change.

Procedure 4.13 Switching to a Stacked Column Chart Type

	Task Description	Mouse/Keyboard
❶	Point anywhere to the **Chart Area** and click the **Right** mouse button.	
❷	From the **Shortcut** menu, select option **Chart Type**.	
	Excel opens the **Chart Type** *dialog box.*	
❸	Select the **Stack Column Chart sub-type**.	
❹	In order to preview your chart now, press and hold down the **Press and hold to view sample** command button.	
❺	When done, release the mouse and click the **OK** command button.	

A **Stacked Chart** compares the absolute contribution of each value
(here monthly number of complaints) to a total across all categories
As you can see in Figure 4.6, the contributions are not dramatically
different. However, some of the data values do not show clearly. For

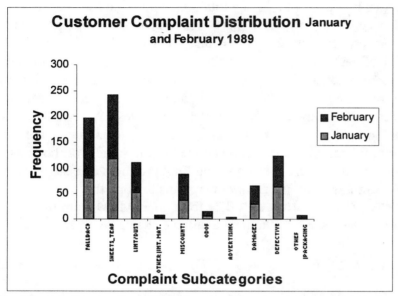

Figure 4.6 Stacked Column Chart for Customer Complaints

example, based on the chart alone, it is very hard to see the difference
between **January** and **February** number of complaints for the category
ADVERTISING. In
such case, you may
wish to use the
**100% Stacked Chart
sub-type**, which
reveals the relative
(percentage)
contributions
(Figure 4.7). This
chart shows clearly
an increase in
customer complaints
for the category
ADVERTISING. We

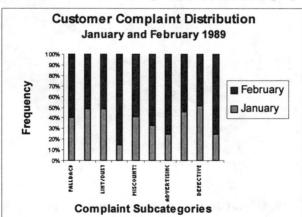

Figure 4.7 100% Stacked Column Chart

Note:
To select the **100%
Stacked Chart sub-
type***, in step* ❸ *of
Procedure 4.13,
click this button:*

now can see that almost all categories feature some level of increased
number of complaints.

Section 4.7 How to Create a Pie Chart in Excel

Qualitative data can also be displayed using a **Pie** chart. A **Pie** chart is particularly useful when you want to look at different categories as a percent of some total. Generating a **Pie** chart is as simple as creating any other chart. There is however one problem. A **Pie** chart, constructed using the **Chart Wizard**, will look well if the number of data points is small and each of the data values represents a significant percent of the total (say, 5% or more). Otherwise, making an informative pie chart may be quite a challenge. Our data set falls in the category of the pie chart's unfriendly cases. We will show why by demonstrating how to create a pie chart for the **January 1989** data. We assume that the chart will be created in the same workbook as the column charts. Figure 4.8 shows the final product. The steps that follow summarize all necessary operations to copy the **SUBCATEGORY** and **NUMBER** data for **January 1989** from the **ClusteredChart** sheet to the next available sheet, say **Sheet3**.

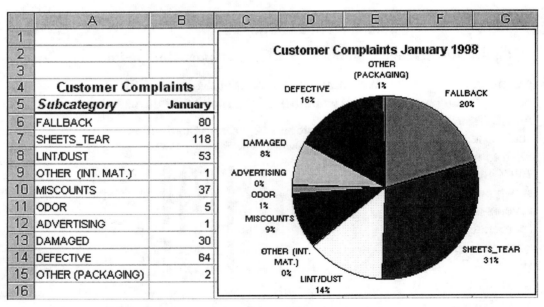

Figure 4.8 Pie Chart for Customer Complaints in January 1989

Procedure 4.14 Preparing Data for the Pie Chart

	Task Description	Mouse/Keyboard
❶	If the **ClusteredChart** sheet is not current, click its **Tab** to switch to it.	
❷	Select the data range representing the **Subcategory** labels and **January** data, including the headings (**C2:D12**).	
❸	Click the **Copy** button.	
❹	Switch to **Sheet3**.	
❺	Select cell **A5**.	
❻	Click the **Paste** button.	
❼	Adjust the width of column **A**.	

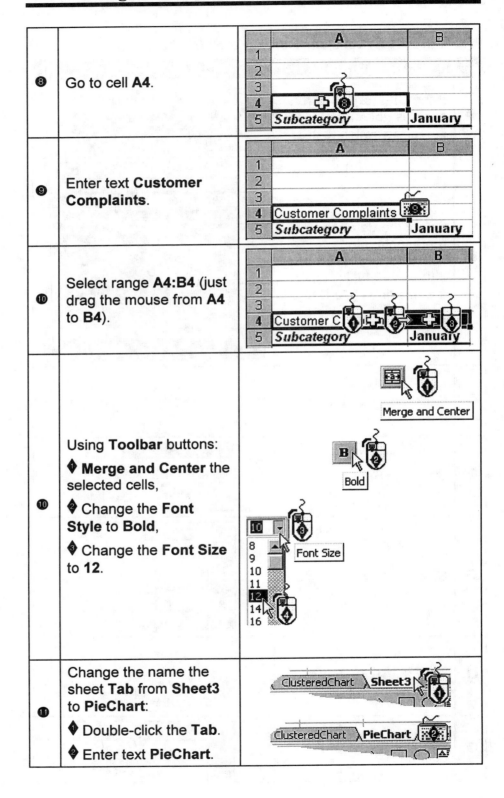

⑧	Go to cell **A4**.
⑨	Enter text **Customer Complaints**.
⑩	Select range **A4:B4** (just drag the mouse from **A4** to **B4**).
⑩	Using **Toolbar** buttons: ◆ **Merge and Center** the selected cells, ◆ Change the **Font Style** to **Bold**, ◆ Change the **Font Size** to **12**.
⑪	Change the name the sheet **Tab** from **Sheet3** to **PieChart**: ◆ Double-click the **Tab**. ◆ Enter text **PieChart**.

The following procedure shows how to create the chart.

Procedure 4.15   Creating a Pie Chart

	Task Description	Mouse/Keyboard
❶	Select the **Source Data** range: ◆ Go to cell **A6**, ◆ Press **END**, **↓**, and **→** while holding down **SHIFT**.		A	B | 1		| 2		| 3		| 4	**Customer Complaints**	| 5	*Subcategory*	**January** | 6	FALLBACK	80 | 7	SHEETS_TEAR	118 | [Shift][End][↓][→]
❷	Click the **Chart Wizard** button.	Chart Wizard
❸	Double-click the **Chart type: Pie** option.	Chart type: Chart sub-type: Column Bar Line Pie XY (Scatter)
❹	Given **Data range:** is **=PieChart!$A$6:$B$15** and orientation is **Series in: Columns**, click the **Next >** command button.	Data range: =PieChart!$A$6:$B$15 Series in: ○ Rows ● Columns Next >
❺	In the **Chart title:** box, type **Customer Complaints January 1989**.	Chart Wizard - Step 3 of 4 - Chart Titles Legend Data Labels Chart title: Customer Complaints January
❻	◆ Click the **Legend Tab**, and ◆ clear the **Show legend Check** box.	Chart Wizard - Step 3 of 4 - Chart Titles Legend Data Labels ☐ Show legend

Note:
Using **SHIFT** *and the navigation keys* (**END ↑ ↓ ← →**), *you can quickly select a range of cells filled with data.*

Note:
Make sure the **Show legend** *check box is empty:*

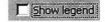

⑦	♦ Click the **Data Labels Tab**, and ♦ click the **Show label and percent Option** button.	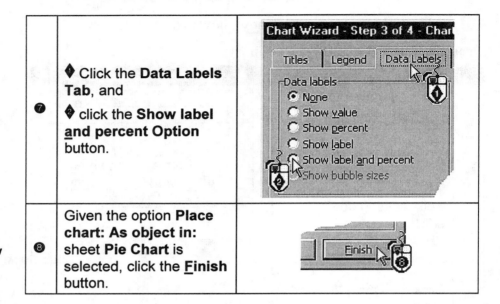
⑧	Given the option **Place chart: As object in:** sheet **Pie Chart** is selected, click the **Finish** button.	

Note:

*Make sure the **Show label and percent Option** button is turned on:*

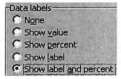

Figure 4.9 shows the resulting chart. As you can see, the data labels are scrambled and pie slices corresponding to the small data

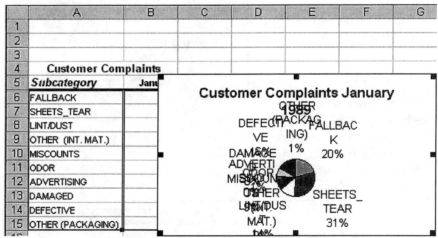

Figure 4.9 Chart Wizard Generated Pie Chart

values are practically invisible. With a help of a few mouse operations you will be able to refine the chart.

To begin your formatting tasks, move and enlarge the chart. The following procedure shows how to do it.

Procedure 4.16 Moving and Enlarging an Embedded Chart

Task Description	Mouse/Keyboard
❶ To move the chart do this: ◆ Place the mouse pointer on the **Chart Area** (near the upper left corner) and hold down the left mouse button. ◆ Drag the chart up and slightly right. ◆ Drop its upper-left corner onto cell **C1**.	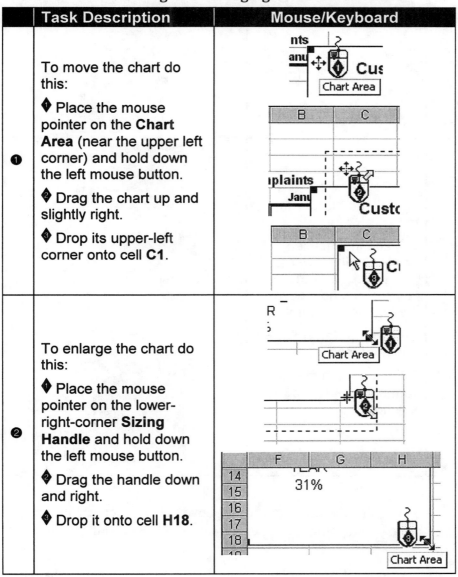
❷ To enlarge the chart do this: ◆ Place the mouse pointer on the lower-right-corner **Sizing Handle** and hold down the left mouse button. ◆ Drag the handle down and right. ◆ Drop it onto cell **H18**.	

Next, adopt the steps of Procedure 4.5 to reduce the **Font Size** of the **Data Labels** to **6** points. *Note*: in Step ❹, type **6** in the **Size** box, instead of selecting the size from the list.

Your chart should already be more readable. It will look even better after you increase the size of the pie itself.

Procedure 4.17 Changing the Size the Pie

	Task Description	Mouse/Keyboard
❶	Click the chart's **Plot Area**.	
❷	To enlarge the pie do this: ◆ Place the mouse pointer on the lower-right-corner **Sizing Handle** and hold down the left mouse button. ◆ Drag the **Handle** down and right, and ◆ Drop it when the pie cannot be enlarged anymore.	

Note:

*It is a good time
to save your
workbook:*

If you have some extra room above the pie, increase its size even more by dragging the upper-left-corner **Sizing Handle** up and left.

Manipulating the pie is somewhat tricky. If you click the pie (inside) for the first time, Excel will select all its slices. Then, if you

*The first click selects
all pie slices.*

*The second click picks
the pointed-to-slice.*

*Drag the slice if you
want to explode it.*

dragged the mouse to the outside, all the slices would get exploded. If you click a particular slice while all the slices are selected, Excel will select that particular slice. In this state, if you dragged the mouse, the selected slice would become exploded.

If some data labels still overlap or are too close, you may need to move some of them manually. As with the pie slices, the first click

Note:
When you are selecting a single label, do not double-click (🖱) the label. Instead, click it once (🖱), wait a second, and then click (🖱) it again.

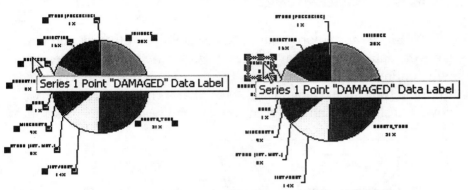

The first click selects all **Data Labels**.   *The first click selects all* **Data Labels**.

selects all the labels. The second one opens the text box with the label pointed at. Now, you can move the box by dragging its border or change some of its properties.

Exercise 4. Create pie charts for the complaint sub category data for **January 1990** and then for **January 1991**. Which do you think is more informative for this data, a pie chart or a column chart?

Section 4.8 Consolidating Data

It seems that both the column chart and the pie chart have too many categories to display and thus they are not quite readable. In other words there is too much information to digest. Although you "fixed" the problems by changing the appearance of the graph, the real problem is that the chart had too many classifications to be effective. An alternative is to create a chart using the complaint **Category** variable instead of the **Subcategory** one. In Excel, this is a three-step process:

▉ Summarize data (calculate automatic subtotals).

▉ Hide detail data.

▉ Generate the chart based on the subtotals.

To implement these steps you will first compute complaint frequency subtotals for each **Category** (for **January 1989**), next hide the detail (**Subcategory**) data and then set up a chart based on the summary (consolidated) data.

Note:
Procedures 4.10 and 4.14 show how to copy and paste data between different sheets.

Since only a portion of the data set will be utilized, you will copy the data to the next available sheet, say to **Sheet4**, and then you will continue processing it there. Figure 4.10 shows the data copied from the **Data** sheet (range **C1:E11**) to **Sheet4** (range **C1:E11**). By now, you should be an expert in copying and pasting spreadsheet data.

Note:
Check step ❿ *of Procedure 4.10 to find out how to adjust the column width.*

	A	B	C
1	*Category*	*Subcategory*	*Number*
2	DISPENSING	FALLBACK	80
3	DISPENSING	SHEETS_TEAR	118
4	FOREIGN MATL.	LINT/DUST	53
5	FOREIGN MATL.	OTHER	1
6	MISCOUNTS	MISCOUNTS	37
7	ODOR	ODOR	5
8	PACKAGING	ADVERTISING	1
9	PACKAGING	DAMAGED	30
10	PACKAGING	DEFECTIVE	64
11	PACKAGING	OTHER	2

|◀ ◀ ▶ ▶| PieChart **Sheet4**

Figure 4.10 Customer Complaint Data for January 1989

Since the data is already sorted by major **Category**, you can compute the **Category** subtotals using the **Data | Subtotals** command.

Procedure 4.18 Generating Compliant Subtotals

	Task Description	Mouse/Keyboard
❶	Copy range **C1:E11** from the **Data** sheet to **Sheet4**.	**Procedure 4.10**
❷	Select any cell of the range **Sheet4! C1:E11**.	A B 1 *Category* *Subcategory* 2 DISPENSING FALLBACK

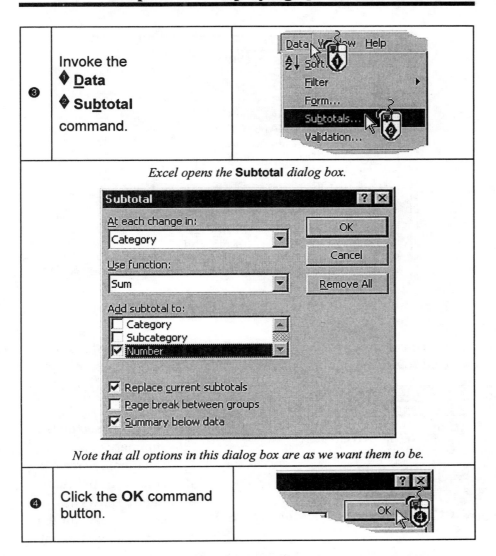

| ③ | Invoke the
♦ **Data**
♦ **Su_b_total**
command. | |

*Excel opens the **Subtotal** dialog box.*

Note that all options in this dialog box are as we want them to be.

| ④ | Click the **OK** command button. | |

Figure 4.11 demonstrates the summarized data. Notice that Excel has inserted extra rows for the subtotals at the end of each **Category** group as well as the grand total at the bottom of the list. It also added the outline control operators at the left that can be used to hide/show different levels of detail. You may wish to explore the controls to see how Excel interprets them.

1 2 3		A	B	C
	1	*Category*	*Subcategory*	*Number*
	2	DISPENSING	FALLBACK	80
	3	DISPENSING	SHEETS_TEAR	118
	4	**DISPENSING Total**		198
	5	FOREIGN MATL.	LINT/DUST	53
	6	FOREIGN MATL.	OTHER	1
	7	**FOREIGN MATL. Total**		54
	8	MISCOUNTS	MISCOUNTS	37
	9	**MISCOUNTS Total**		37
	10	ODOR	ODOR	5
	11	**ODOR Total**		5
	12	PACKAGING	ADVERTISING	1
	13	PACKAGING	DAMAGED	30
	14	PACKAGING	DEFECTIVE	64
	15	PACKAGING	OTHER	2
	16	**PACKAGING Total**		97
	17	**Grand Total**		391

Figure 4.11 Summary of Complaint Data for January 1989.

To continue, you need to click the level ☒ control which will cause all rows containing the third level detail data (subcategories) to be hidden (Figure 4.12).

1 2 3		A	B	C
	1	*Category*	*Subcategory*	*Number*
⊞	4	**DISPENSING Total**		198
⊞	7	**FOREIGN MATL. Total**		54
⊞	9	**MISCOUNTS Total**		37
⊞	11	**ODOR Total**		5
⊞	16	**PACKAGING Total**		97
	17	**Grand Total**		391

Figure 4.12 Subtotals with Hidden Detail Data.

Based on the summary data, you can quickly create a column or pie chart. Figure shows a column chart for the summary complaint data. The steps that follow explain how to create this chart.

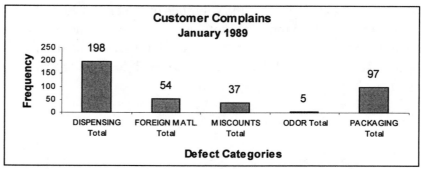

Figure 4.13 Column Chart for Major Category Defects

Procedure 4.19   Creating a Column Chart Based on Subtotal Report

	Task Description	Mouse/Keyboard
❶	Select two non-adjacent ranges **A4, A7, A9, A11, A16; C4, C7, C9, C11, C16**: ◆ Click **A4**, ❷ Hold down SHIFT and click **A16**, ❸ Hold down CTRL and click **C4**, ◆ Hold down SHIFT and click **C16**,	(spreadsheet image showing columns A, B, C with rows: Category / Subcategory / Number; DISPEN... Total / Ctrl / 198; FOREIGN MATL. Total / 54; MISCOUNTS Total / 37; ODOR Total / 5; PACKAGIN...T Shift / Shift / 97; Grand Total / 391)
❷	Run the **Chart Wizard** and format the chart.	Procedures 4.1, 4.2, 4.3, etc.

Note: To select non-adjacent ranges mark the beginning of each separate range with

Ctrl

and the end with

Shift

Exercise 5. Create column charts for the major category classifications for **January 1990** and **January 1991**.

Section 4.9 Pareto Diagrams

The bar chart just created for the major defect categories (Figure 4.13) is an improvement over the previous attempts, but it still leaves something to be desired. Remember that management is trying to identify the primary cause of customer complaints and that while the information is available in our column chart, it is not emphasized. Data of this type is better displayed with the columns arranged in descending order of frequency. We can accomplish this in Excel by sorting the chart's data on the **Number** of complaints, in descending order.

Procedure 4.20 Sorting the Subtotals by Number

	Task Description	Mouse/Keyboard
❶	Select a **Number** value cell.	(spreadsheet image showing columns A, B, C with rows: Category / Subcategory / Number; DISPENSING Total / 198; PACKAGING Total / 97)

❷	On the **Standard Toolbar**, click the **Sorting Descending** button.	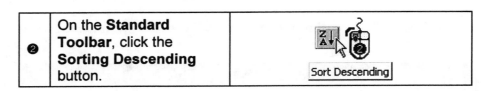

Excel rearranges the data and updates the chart (Figure 4.14). This column chart gives management more insight into where efforts to reduce customer complaints should be directed. If the data are to be incorporated into a presentation of any type, this chart will be the most effective in conveying the information.

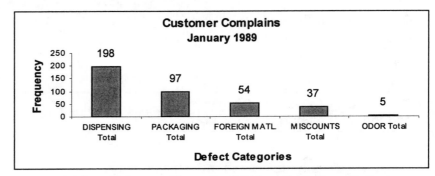

Figure 4.14   Column Chart for Major Category Defects Sorted in Descending Order

What you have just created is closely related to a tool from **Statistical Process Control** known as a ***Pareto Diagram***. A Pareto Diagram is often used to identify trouble spots in defect and complaint data. The Pareto Diagram is a column chart with the column-bars arranged in order of decreasing frequency, but it adds a line that plots the *cumulative frequency* of the categories (not present in our chart). This not only tells management which categories are the largest, but it gives information about what percent of the total problems are accounted for by the largest categories.

Exercise 6. Create a Pareto chart for **January 1990** and compare it to the one for **January 1989**. How are they the same? How are they different?

Section 4.10 Looking at the Data for Different Time Periods

So far you have looked at the data in the time frame that it was collected, a month at a time. With so much data available, it would not make sense to make decisions based on any single month. In fact there is no real reason to believe that all of the months are similar. When data is available for many different time periods it is usually intelligent to look at the data for these different periods in several different ways. For example, the company might be interested in looking at the trend of complaints over an entire year to see if there is any information to be obtained.

Since the **Data | Subtotals** command allows you to sum the data over the categories, all that is required to create a column chart that looks at monthly data is to execute this command for all the **1989** data and then set up the chart based on the summary data.

On the **Data** sheet, copy the data for **1989** (range **A1:E121**) and paste it to **Sheet5**. While the range is still selected, invoke the **Data | Subtotals** command. Excel should automatically select **Month** in the

At each change in box and **Number** in **Add subtotal to** box. Press ENTER or click the **OK** button to accept the settings. On the left-hand side outline, click the level ② control. Finally, following the steps from Procedure 4.1, create a column chart. For the chart data, select range **E12:E133** only. This will force Excel to define **the X-axis** labels using integers **1** through **12** (coincidentally to represent months **January**

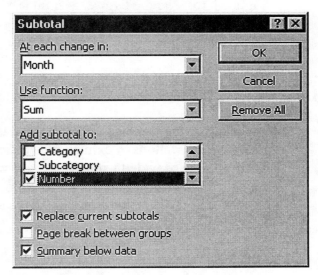

through **December**). Do not sort the data this time, as it would disturb the chronological order. The result should look something like the one in Figure 4.15.

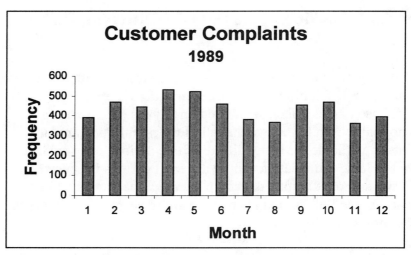

Figure 4.15 Monthly Customer Complaint Data

Exercise 7. Do you notice any interesting features about the number of complaints over the year? Can you offer any explanations for why this might occur?

Exercise 8. Create a column chart that shows the total defects by month over the three-year period. Compare this graph to the one in Figure 4.15. Are the two years similar? If not, how are they different?

Section 4.11 Investigative Exercises

In the following exercises you are asked to use the skills introduced in the previous sections to extract information from the complaint data file. You are provided with space to answer the questions and paste in graphical output from the program. If you do not have access to a printer, you can sketch the graphs on the axes provided.

1. a) Construct a column chart of the packaging sub categories complaints for January, 1990.

    b) Construct a pie chart for the same data.

    c) What packaging sub category contributes the most complaints? Which is second? Which graph shows this better? Why?

2. a) Create a column chart for packaging sub categories for January 1991.

    b) Do the same thing for October 1990.

    c) Do the distribution of defects appear to be the same for the two graphs that display January data? Would you expect them to be?

    d) Is the distribution for October similar to these two? Would you expect it to be?

3. a) Create pie charts for October 1990 and January 1991.

    b) Is it easier to make comparisons using bar charts or pie charts?

4. a) Create a clustered column chart for October 1990 and January 1991.

    b) Create a stacked column chart for the same data.

c) Are comparisons like this easier using a single graph or three different graphs? What things do you have to be careful of when you use three different graphs?

5. a) Calculate the total number of packaging complaints for each month of 1989.

b) Create a column chart that displays this data.

c) Examine the column chart you created in part b. What do you notice about the number of packaging complaints over the year? Is it constant? If not, what do you see happening?

d) Create a pie chart for this data. Can you answer part c using the pie chart? Why or why not?

6. a) Repeat exercise 4, parts a and b for the packaging data from 1990 and then for 1991.

b) Make a clustered column chart of the packaging data from the three years.

c) Compare the behavior of the number of packaging complaints over the years. What do you observe?

7. a) Look at the total number of complaints by month for 1990 and 1991.

b) Make column charts of the monthly totals for these two years.

   c) Compare these graphs to each other and to the graph for 1989 that you created earlier in the chapter. Are there any noticeable trends or patterns? What differences or similarities do you see among the years?

   d) What months or seasons generally have the most complaints? The fewest?

   e) Can you offer an explanation for this?

---

8. Compare the monthly bar charts for packaging defects to the ones for total defects. Are they similar ? Would you expect them to be related to each other? Why or why not?

---

9. a) Create pie charts for total complaints using the variable *Category* for each of the years.

   b) Have there been any major shifts in the relative frequency of the types of complaints over the past three years?

   c) Display the same information on a clustered bar chart and a stacked bar chart.

   d) Which graph do you think is more informative? Why?

   e) Do you get the same information from these display that you get from the pie charts? Which type do you think will give the company a better picture of their complaint history? Why?

10. a) Suppose you wanted to compare the distribution of complaints (by category) on a monthly basis to see if there are any similarities. What type of graph(s) would you use? Why?

b) Do this for one of the months. What is your conclusion?

11. The management of this company asks you for your assessment of their complaint data. In particular they would like to know where their biggest problems are and what the trends seem to be over the past three years. Create any additional graphs that will help you answer their questions and prepare a report for them offering your analysis and your suggestions.

Chapter 5 "Golf Ball Design"

Displaying Quantitative Data

Section 5.1 Overview

Statistical Objectives: After reading this chapter and doing the exercises you will:

- Know the function of a histogram.
- Know how to describe the shape of a distribution from the histogram.
- Know the effects of changing class interval widths.
- Know the effects of changing the number of class intervals.
- Know how to detect outliers using histograms.
- Know how to compare variables using histograms.

Section 5.2 Problem Statement

All companies are currently facing increasing competition at the national and international level. In the face of this competition American manufacturers are moving to a focus on *quality*. This is not a simple change but rather encompasses all aspects of the business and all employees of the company. Managers are learning to listen to the creative ideas of their employees, are breaking down the hierarchical management layers and monitoring their production processes in order to prevent problems ahead of time. In doing so, more and more data is being collected on every aspect of the business. This includes data on such things as customer complaints as well as data on how closely the product matches with the target design. It means all parts of the organization must work together as a team to achieve Total Quality Management (TQM).

In this chapter you will look at how a large manufacturer of golf balls is incorporating some of these issues. Many of you enjoy taking to the golf course at the first hint of good weather. In some parts of the country you can golf every day of the year while others must patiently wait for the snow to melt. Next time you hit the golf ball, think about what type of decisions went into the production of that golf ball in order to make you look good on the golf course -- after all we all want our balls to fly a great distance when we hit them! What can the manufacturer do which ultimately will affect how far the ball flies?

In this particular case the company is trying to evaluate two different ball designs. Which of them will repeatedly perform the best on the golf course? Anyone who has hit a golf ball knows that in addition to the particular ball design, there are many factors that could influence how far the ball flies; for example the expertise level of the golfer, the wind speed, etc.

In fact, there are many factors that effect how far the ball flies and in order to remain competitive, most golf ball manufacturers are constantly creating and testing new ball designs. Such things as ball size, weight, and dimple pattern are all within the control of the manufacturer and are called **internal factors**.

Figure 5.1 A Golf Ball

Such things as the expertise level of the golfer, head speed, launch angle, wind speed, wind direction, relative humidity and temperature also effect the distance the ball flies but are not typically within the control of the manufacturer! These are called **external factors.**

The company is primarily interested in measuring how far the ball carries and the total distance the ball travels. It should be noted that *Carry* is how far the ball has gone from the point where it was hit to where it lands. *TotDist* (total distance) also includes the distance that the ball has rolled after it hits the ground. Typically, total distance is greater than carry. Think about how these characteristics can be measured.

This company has a testing site in Florida and balls are placed into machines that hit the balls at a pre-specified speed and launch angle. Many balls are hit by the machine within a short period of time so that factors such as wind speed, wind direction, relative humidity, and temperature are not fluctuating very much. Measurements are taken on these factors to be sure that they are basically constant throughout the test. Thus the external factors are being controlled.

For a given test, the balls used are all the same model number with the same dimple design and they should be the same weight and size (although these are also measured). Thus, the internal factors are being controlled.   The company has attempted to create a situation where all the factors that we listed above, both internal and external factors, are held constant in order to study the characteristics of interest: *carry and total distance*.

NOW the big problem is to measure how far the ball carries! The total distance is fairly easy to measure as we have the ball in its final resting spot as a marker. In order to measure carry we need to know where the ball lands -- not where it finally ends up after rolling a bit. SO -- the company hires people to stand in the fields and place markers where the balls actually hit the ground.   This is true! If you think you have a headache reading this, think about how those field researchers feel!

The data you will be analyzing in this chapter contains information on most of the factors that have just been discussed. Two different ball designs are being compared and the data were taken during three different time periods.

Section 5.3 Characteristics of the Data Set

FILENAME:	Ch05Dat.xls	An Excel Workbook
SIZE:	COLUMNS	14
	ROWS	72

The first 7 rows of the actual data file are shown in Figure 5.2.

	A	B	C	D	E	F	G	H	I	J	K	L	M	N
1	Ball	Model	Size1	Size2	Size3	Wgt	Dw	Dd	Head	Temp	Carry	TotDist	ODate	OTime
2	1	M1	81	81	82	45.3	0.145	0.011	686	77	257	270	20-Aug	8:15
3	2	M1	83	83	84	45.2	0.151	0.011	688	77	255	267	20-Aug	8:15
4	3	M1	81	82	84	45.2	0.145	0.010	687	77	256	267	20-Aug	8:15
5	4	M1	81	81	83	45.3	0.144	0.012	688	77	255	271	20-Aug	8:15
6	5	M1	83	81	82	45.5	0.146	0.011	687	77	255	268	20-Aug	8:15
7	6	M1	83	83	82	45.3	0.156	0.011	687	77	256	267	20-Aug	8:15
8	7	M1	81	81	82	45.2	0.149	0.011	687	77	255	264	20-Aug	8:15

Ch05Dat.xls — Data / Sheet2 / Sheet3 / Sheet4

Figure 5.2 A Fragment of the Ch05Dat Data Set

Notes on the data set:

1. The variable *Ball* keeps track of the observation number and goes from 1 to 72.

2. The variable *Model* indicates which of the two ball designs is used for that observation. There are two different Models in this data set:
    M1:   Ball 1-12, 25-36, 49-60
    M2:   Ball 13-24, 37-48, 61-72

3. The variables *Size1, Size2,* and *Size3* are size measurements. They indicate the measurement of the ball around the ball's equator, and at two other points. All three measurements should be very close if the ball is spherical.

4. The variable *Wgt* indicates the weight of the ball.
5. The variables *Dw* and *Dd* are measurements of the dimples. *Dw* indicates the dimple width and *Dd* indicates the depth.

6. The variable *Head* is the speed on the ball when it is hit.

7. The variable *Temp* is the air temperature measured in degrees Fahrenheit.

8. The variable *Carry* indicates the distance from the point the ball was hit to the point where it hit the ground, measured in yards.

9. The variable *TotDist* indicates the total distance the ball has traveled. It is equal to the variable Carry plus the distance the ball rolled after hitting the ground.

10. The *ODate* is 8/20 for all 72 observations.

11. The variable *OTime* shows the time of day when the observation was recorded. There are 3 different time periods:
8:15 AM -- Ball 1-24
8:45 AM -- Ball 25-48
9:15 AM -- Ball 49-72

Start Excel and open the **Ch05Dat.xls** data file. Refer to Procedure 3.1 (page 33) for help on how to open a workbook. It is recommended that you use a copy of your original diskette. From Procedure 2.2 (page 22), you can learn how to duplicate the diskette.

Open

Section 5.4 How to Create a Histogram in Excel

In this section you will learn how to create frequency histograms, relative frequency histograms and manage graphs.

Section 5.4.1 Creating a Frequency Histogram

This section focuses on the basics of creating a frequency histogram using Excel. Subsequent sections cover some of the more advanced features available in Excel, which pertain to histograms.

Suppose you want to generate a histogram for the variable *Carry* for all **72** observations, using the tools provided by the **Data Analysis ToolPak**. The values of this variable reside in the range **K2:K73**. It will be more convenient and meaningful to work with Excel commands and function, if we name this range as *Carry*. For example, if you are requested to provide the input range for this variable, typing *Carry* is easier and more informative than **K2:K73**. In fact, the following procedure shows how to name all the variables, so that later they can be referred to by names rather than by their "mysterious" ranges.

Note:
Make sure the **Tools** *menu contains the* **Data Analysis** *option. If not, you may have to run the* **MS Office Setup** *program to include the* **Analysis ToolPak** *and then execute the* **Tools | Add-Ins** *command.*

Procedure 5.1 Naming All Variables of the Data Set Ch05Dat

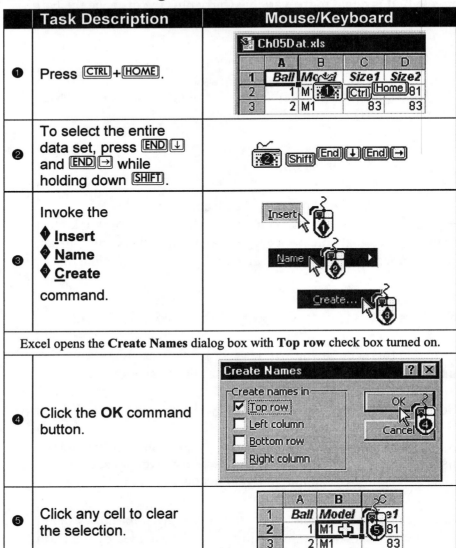

	Task Description	Mouse/Keyboard
❶	Press $\boxed{\text{CTRL}}$ + $\boxed{\text{HOME}}$.	
❷	To select the entire data set, press $\boxed{\text{END}}$ $\boxed{\downarrow}$ and $\boxed{\text{END}}$ $\boxed{\rightarrow}$ while holding down $\boxed{\text{SHIFT}}$.	
❸	Invoke the ◆ **Insert** ◆ **Name** ◆ **Create** command.	
	Excel opens the **Create Names** dialog box with **Top row** check box turned on.	
❹	Click the **OK** command button.	
❺	Click any cell to clear the selection.	

Note:
In Excel, pressing $\boxed{\text{CTRL}}$ + $\boxed{\text{HOME}}$ *brings you to the beginning of the current worksheet (**A1**).*

Note:
*By showing the **Top row** check box turned on, Excel indicates that it is ready to use the top row labels to name the data located right below the labels.*

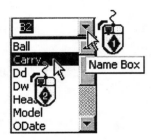

Hint:
The workbook names can be quickly accessed from the name box, located on the **Formula Bar**. For example, in order to select the range named as *Carry*, click the **Name Box Drop-down** arrow and select the name from the list.

Now, let us move on to generating a frequency histogram for the variable *Carry*.

Procedure 5.2   Generating a Histogram Using the Data Analysis ToolPak

	Task Description	Mouse/Keyboard
❶	Invoke the ◆ **Tools** ◆ **Data Analysis** command.	
	Excel opens the **Data Analysis** *dialog box.*	
❷	Double-click option **Histogram**.	
	Excel opens the **Histogram** *dialog box.*	
❸	◆ In the **Input Range** Text box, type **Carry**, ◆ Click the **New Worksheet Ply:** text box, ◆ Type the name of the new sheet CarryHistogram, ◆ Turn on the **Chart Output** check box, and ◆ Click the **OK** command button.	

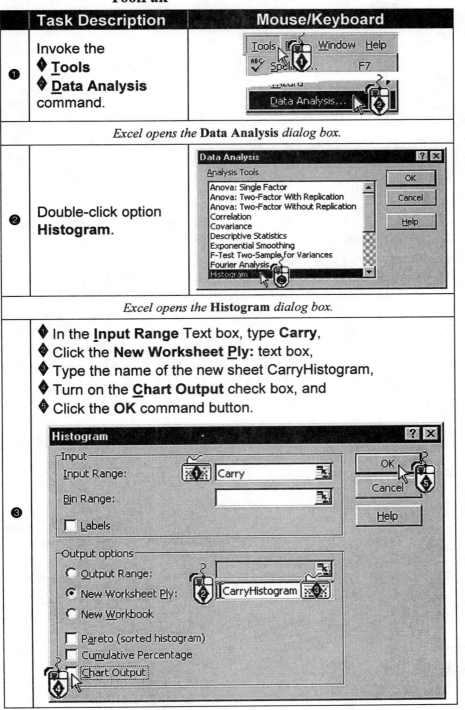

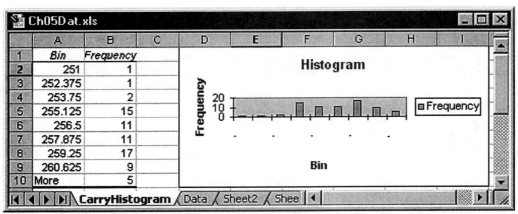

Figure 5.3   Frequency Distribution and Histogram Generated by the Data Analysis Histogram tool.

Figure 5.3 shows the result. Excel has created a histogram by first setting up the default class intervals (bin) and then counting the number of *Carry* observations (frequency) in each of the intervals. For example, Excel found **1** observation up to **251** yards, **15** observations above **253.75** yards and below or equal to **255.125** yards. The More row shows the number of observations above the highest interval (*Bin*) limit (there are **5** values above **260.625**).

Note:
If you do not define the **Bin Range***, Excel will create it for you.*

Based on the **Bin** and **Frequency** values, Excel has generated the default column chart. Note that without activating the **Chart Output** option, Excel would only generate the table histogram.

In Chapter 4, you learned how to create, format, and edit column charts. Now, it is time to use your skills and bring the histogram to approximately what is shown in Figure 5.4.

Hints:

To eliminate the **Legend** *box, click it and press* DEL.

To change a **Chart**, **Category**, *or* **Value** *Title, click the* **Title** *box and type new text.*

To show one-decimal-place format for the **Category Axis Numbers**, *select the* **Bin Data Source** *and click the* **Increase Decimal** *button.*

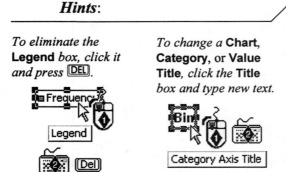

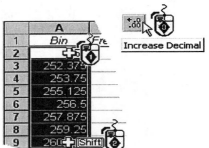

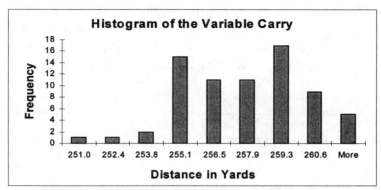

Figure 5.4 Histogram of *Carry* for all Observations

When done, you may wish to print the chart alone or together with the table.

Exercise 1. What does the histogram tell you about the distribution of how far the balls carry? How far do most of the balls carry?

Section 5.4.2 The Relative and Cumulative Frequency Distribution

Notice that the vertical axis of the histogram (Figure 5.4) shows the frequency or raw count for each of the columns in the graph. Sometimes it is more helpful to have the relative frequency or percent displayed on the vertical axis. Assessment of some probabilities can be greatly simplified with a help of a cumulative distribution. In this section, you will generate relative and cumulative frequency distributions based on your own class interval (**Bin**) setup. This time, you will use Excel common functions and commands (rather than those included into the **Data Analysis Toolpak**).

Rules for creating the class intervals are quite simple. The number of the intervals is expected to be <u>close</u> to the square root of the sample size but not less than **5** and usually not greater than **20**. The width of each interval should be <u>approximately</u> equal to the difference between the sample maximum and minimum (sample range) divided by the number of the intervals. As you will see, the default class intervals produced by the **Tools | Data Analysis | Histogram** command adhere to these rules. A small departure from the rules is allowable.

Note:
To print the frequency table and chart, select a range covering both the objects and then invoke the **File | Print** *command. In the* **Print** *dialog box, click the* **Selection** *radio button followed by the* **OK** *button.*

Before you can apply the rules, you will need to process the variable *Carry* in order to find its size, maximum, minimum, range, and to perform other calculations. Figure 5.5 shows how to obtain the information necessary for creating the class intervals. The left part

Note:
To right-align labels in range **A14:A19**, *select the range and then click the* **Align Right** *icon:*

Align Right

	A	B
14	n	=COUNT(Carry)
15	Min	=MIN(Carry)
16	Max	=MAX(Carry)
17	Range	=B16-B15
18	m	=INT(SQRT(B14))
19	w	=B17/B18

	A	B
14	n	72
15	Min	251
16	Max	262
17	Range	11
18	m	8
19	w	1.375

Figure 5.5   The Reference Information for the Class Interval Setup

specifies what you enter, the right part shows the outcomes.
Note:

- Function **COUNT** determines the size (**n**) of the *Carry* sample (it counts the number of non-blank, numeric cells within the range named *Carry*).
- Functions **MIN** and **MAX** calculate the sample's minimum and maximum, respectively. They are used to find out the sample's range in cell **B17**.
- Cell **B18** contains the function **SQRT** nested in the function **INT**. Excel first calculates the square root (**SQRT**) of the value stored in cell **B14** and then takes the integer (**INT**) part of the root (the fraction gets truncated). The resulting value represents the approximate number, **m**, of the class intervals.
- Cell **B19** stores a rough estimate of the width, **w**, of each class interval. Notice that **m** and **w** are somewhat inversely proportional. If **w** increases, **m** should decrease and vice versa.

Note:
Some of the procedures presented in our workbook have been automated in form of Excel macro commands and functions. Such procedures are accompanied by this icon:

Now, if necessary, switch to the sheet **CarryHistogram** and enter the labels and formulas as shown in Figure 5.5. If you are very busy at the moment, you may wish to get this job done with a very little effort on your behalf. The following procedure shows how.

Procedure 5.3 Running the GetInfoForBinSetup Macro

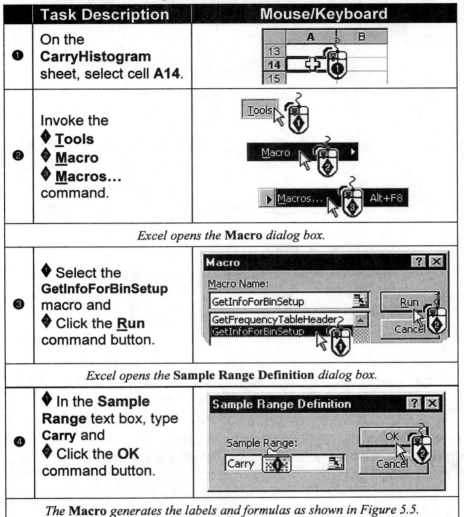

*The **Macro** generates the labels and formulas as shown in Figure 5.5.*

Based on the information obtained manually or through the macro, you will form **7** class intervals (close to **m = 8**). Each of the **5** internal intervals has a width of **2** (close to **w = 1.375**). Starting from the sample minimum (**Min = 251**) your intervals (bins) can be defined as follows:

(up to 251], (251,253], (253,255], (255,257], (257, 259], (259, 261], (more than 261)

Note:
A parenthesis '(' and bracket ']' define here exclusive and inclusive limits, respectively. For example, the sample frequency of the second interval is the number of sample values greater than 251 (exclusive) but less than or equal to 253 (inclusive).

Excel expects the limits of the class intervals (the bin range) to be defined in a column-like range. Figure 5.6 shows the intervals along with headings for columns reserved for the frequencies. Note that the bin range forms an arithmetic sequence. Therefore, instead of

	A	B	C	D
14	n	72		
15	Min	251		
16	Max	262		
17	Range	11		
18	m	8		
19	w	1.375		
20				
21	Bin	Frequency	Relative Frequency	Cumulative Frequency
22	251			
23	253			
24	255			
25	257			
26	259			
27	261			

Note:
*In order to automatically generate this frequency table header, select cell **A21** and run the **GetFrequencyTableHeader** macro (**T**ools | **M**acro | **M**acros | GetFrequencyTableHeader | **R**un).*

Figure 5.6 Class (Bin) Range Setup.

entering all the numbers manually, you can enter the first two numbers in **A22** and **A23**, next select the cells, and then drag the fill handle down to cell **A27**. Here is a detail instruction.

Procedure 5.4 Generating Class Interval Limits (Bins)

	Task Description	Mouse/Keyboard
❶	Enter **251** in **A22** and **253** in **A23**.	
❷	Select the two cells.	

③	Place the mouse pointer on the **Fill Handle** and hold down the left button.	
④	Drag the handle down.	
⑤	Drop it in cell **A27** (when you reach **261**).	
	Note: Keep the selection "alive" (highlighted). Do not clear it.	

The **A22:A27** range constitutes the class intervals. In generating the frequency distribution it will be easier to refer to it by name, say **Bin**, which is why your next task is to name this range as **Bin**.

Procedure 5.5 Naming a Range Using the Name Box

	Task Description	Mouse/Keyboard
①	With the **A22:A27** range still selected, click the **Name Box**, located on the left side of the **Formula Bar**.	

Note:
An alternative way to name this range would be this:
*Select range **A21:A27** (along with name Bin in **A21**) and then invoke the **Insert | Name | Create | OK** command..*

Note:
*The **Name Box** can be used for both creating and selecting names (one at a time).*

❷	Right away, type **Bin** and press ENTER.	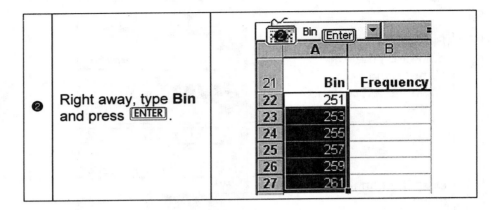

You are about to calculate the frequencies for the sample range named **Carry** (variable *Carry*) and for the class intervals range named **Bin**, using the **Frequency** function. This function returns multiple values in a selected range of cells, which is why it is called an array function.

Procedure 5.6   Generating a Frequency Distribution Using the Frequency Function.

Note:
*In selecting the output range for the **Frequency** function, include one extra cell, beyond the last class limit, unless the limit it greater than the sample **Maximum**.*

Note:
*During development of a function or formula, the range **Name Box** is replaced by the **Function Box**.*

	Task Description	Mouse/Keyboard
❶	Select the frequency distribution outcome range (**B22:B28**).	
❷		

*Excel opens the **Function Assistant** box, suggesting the most recently used function (in our case it is the **Sum** function).*

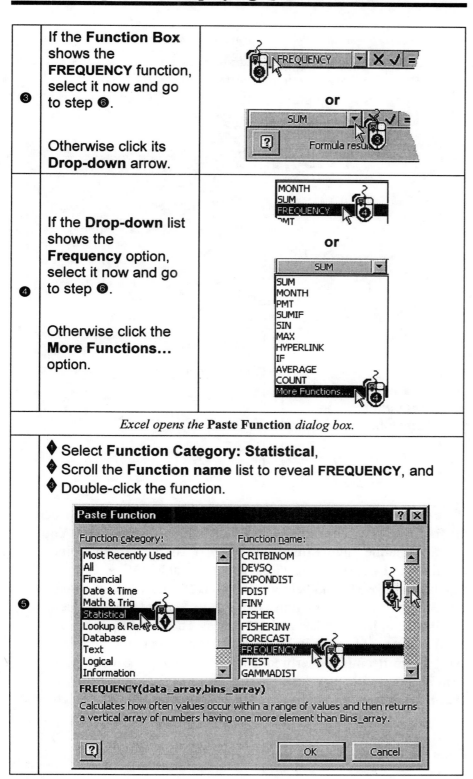

❸	If the **Function Box** shows the **FREQUENCY** function, select it now and go to step ❻. Otherwise click its **Drop-down** arrow.	
❹	If the **Drop-down** list shows the **Frequency** option, select it now and go to step ❻. Otherwise click the **More Functions...** option.	

Excel opens the **Paste Function** *dialog box.*

❺	◆ Select **Function Category: Statistical**, ◆ Scroll the **Function name** list to reveal **FREQUENCY**, and ◆ Double-click the function.

Excel inserts the function onto the **Formula Bar** *and re-opens* **the Function Assistant**.

❻

◆ Click the **Data_array** input box,
◆ Type **Carry** and press **Tab**,
◆ In the **Bins_array** input box, type **Bin**, press ENTER while holding down both SHIFT and CTRL.

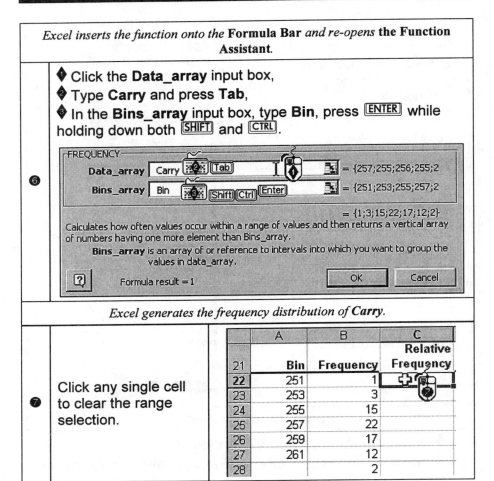

Excel generates the frequency distribution of **Carry**.

Note: *The same frequency could be obtained via the* **Tools | Data Analysis | Histogram** *command.*

❼ Click any single cell to clear the range selection.

	A	B	C
21	**Bin**	**Frequency**	**Relative Frequency**
22	251	1	
23	253	3	
24	255	15	
25	257	22	
26	259	17	
27	261	12	
28		2	

The output range of the function is **B22:B28**, which is where the frequencies associated with the class intervals are defined. You have created the sample frequency of *Carry*. Excel has determined that there is **1** *Carry* distance less than or equal to **251** yards, **3** distances above **251** and below or equal to **253**, etc. Notice also that cell **B28** reveals **2** observations above **261**. Most of the distances (**22** out of **72**) fall between **255** and **257** yards. The extreme (low and high) distances are quite unlikely, and those close to the middle class exhibit significantly higher frequency or likelihood. **Such a shape of data is referred to as normal. Or, one can say that the *Carry* distances show a normal pattern (see Chapter 7 for more details).**

Compare this frequency with the default one (produced by the **Tools | Data Analysis | Histogram** command). As one could expect, due to a smaller number of class intervals (and thus larger interval width), these frequencies are mostly higher.

All the frequencies should add up to **72** (the sample size). To check if this is the case, place the cell pointer in **B29** and click the **Auto Sum** icon.

To define the relative frequencies, each of the absolute frequencies must be divided by the total (the sample size). For example, the relative frequency of the first class interval is **1/72**, the second **3/72**, the third **5/72**, etc.

Procedure 5.7   Generating Relative and Cumulative Frequency Distribution

	Task Description	Mouse/Keyboard
❶	Name the Frequency output range (**B22:B28**) as **Frequency**.	Procedure 5.5
❷	Name the cell containing the size of the *Carry* variable (**B14**) as **n**.	Procedure 5.5
❸	Select the Relative Frequency range (**C22:C28**).	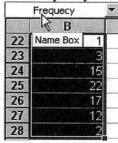
❹	Type this formula: **=Frequency/n** *Note*: **DO NOT** press ENTER !	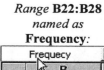
❺	Hold down CTRL and press ENTER .	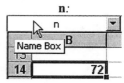

Range **B22:B28** *named as* **Frequency***:*

	B	
Frequecy ▼		
22	Name Box	1
23		3
24		15
25		22
26		17
27		12
28		2

Cell **B14** *named as* **n***:*

n ▼	B
Name Box	
14	72

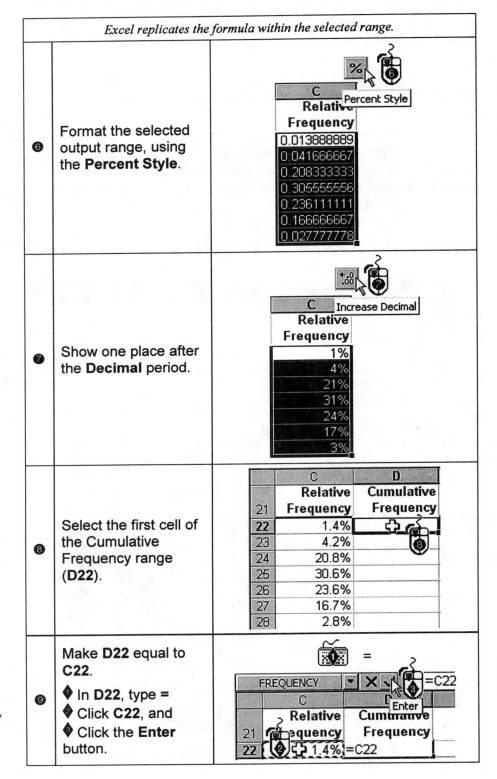

Excel replicates the formula within the selected range.

⑥	Format the selected output range, using the **Percent Style**.

⑦	Show one place after the **Decimal** period.

⑧	Select the first cell of the Cumulative Frequency range (**D22**).

⑨	Make **D22** equal to **C22**. ◆ In **D22**, type **=** ❷ Click **C22**, and ◆ Click the **Enter** button.

Note:
The CTRL + ENTER *operation produces an entry in each cell of a selected range. An alternative solution for the Relative Frequency formula would be to enter =Frequency/n or =B22/$B$14 into cell C22, then copy this formula and paste it onto range C23:C28.*

Note:
For the first class interval (Bin), the **Cumulative Frequency** *is the same as the* **Relative Frequency**.

⑩	Select cell **D23**.	

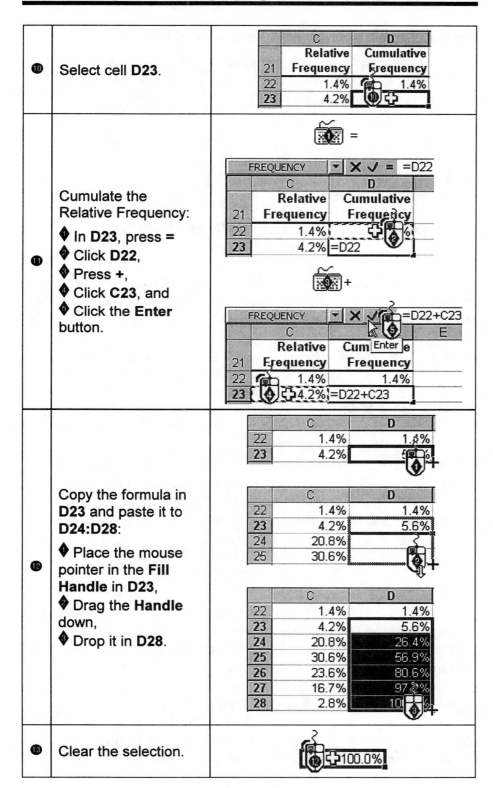

⑪	Cumulate the Relative Frequency: ◆ In **D23**, press **=** ◆ Click **D22**, ◆ Press **+**, ◆ Click **C23**, and ◆ Click the **Enter** button.

⑫	Copy the formula in **D23** and paste it to **D24:D28**: ◆ Place the mouse pointer in the **Fill Handle** in **D23**, ◆ Drag the **Handle** down, ◆ Drop it in **D28**.

⑬	Clear the selection.

Note:
The last
Cumulative
Frequency
value must be
100%.

Here is what your frequency distributions should look like.

	A	B	C	D
			Relative	**Cumulative**
21	**Bin**	**Frequency**	**Frequency**	**Frequency**
22	251	1	1.4%	1.4%
23	253	3	4.2%	5.6%
24	255	15	20.8%	26.4%
25	257	22	30.6%	56.9%
26	259	17	23.6%	80.6%
27	261	12	16.7%	97.2%
28		2	2.8%	100.0%

**Figure 5.7   Relative and Cumulative Frequency
Distribution of *Carry*.**

The **Cumulative Frequency** distribution is very helpful in assessing probabilities of varies outcomes of the variable ***Carry***. Here are a few examples:

P(*Carry* ≤ 253) = 5.6%:   The probability of the golf ball to **carry** a distance less than or equal to **253** yards;

P(*Carry* ≤ 259) = 80.6%:   The probability of the golf ball to **carry** a distance less than or equal to **259** yards;

Note:
$P(X > a) = 1 - P(X \le a)$

P(*Carry* > 259) = 19.4%:   The probability of the golf ball to **carry** a distance longer than **259** yards (**100%-80.6%**);

Note:
$P(a < X \le b) =$
$P(X \le b) - P(X \le a)$

P(253 < *Carry* ≤ 259) = 75.0%:   The probability of the golf ball to **carry** a distance above **253** but less then or equal to yards (**80.6%-5.6%**);

Figure 5.4 shows the frequency histogram of ***Carry*** generated with a help of the **Tools | Data Analysis | Histogram** command. You now have an opportunity to create a similar histogram entirely on your own.

Exercise 2. Based on the **Relative Frequency** distribution, generate the relative frequency histogram using the **Chart Wizard**. Use the **Bin** range (**A22:A28**) for the **Category (X) Axis** and the **Relative Frequency** range (**C22:C28**) for the **Value (Y) Axis**.

Hint:

Procedure 5.8   Creating a Chart Based on Non-adjacent Data Source Ranges

	Task Description	Mouse/Keyboard
❶	Select the **Value (Y) Axis** range (**C22:C28**): ◆ Click **A22**, ◆ Shift-click **A28**.	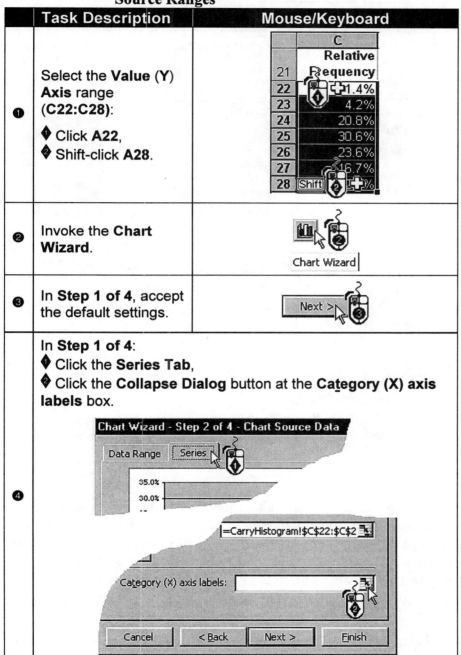
❷	Invoke the **Chart Wizard**.	
❸	In **Step 1 of 4**, accept the default settings.	
❹	In **Step 1 of 4**: ◆ Click the **Series Tab**, ◆ Click the **Collapse Dialog** button at the **Category (X) axis labels** box.	

Note:
You may also type the **Category** *axis data range* (**A22:A28**) *right in the* **Category (X) axis labels** *box. If you do, you will then be able to skip steps* ❺ *and* ❻.

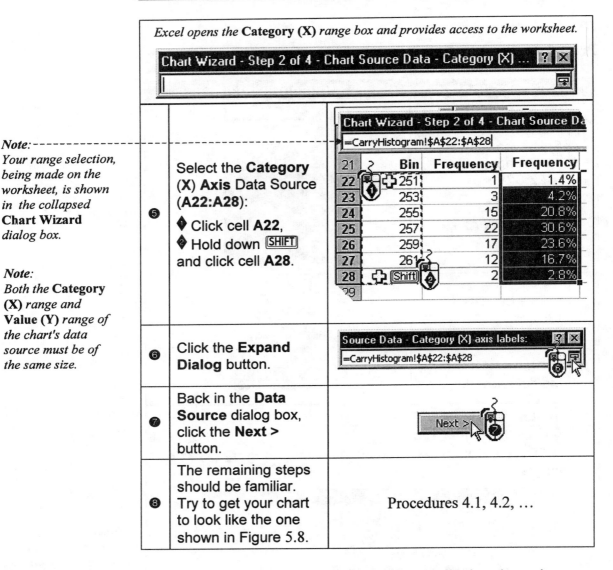

*Excel opens the **Category (X)** range box and provides access to the worksheet.*

Chart Wizard - Step 2 of 4 - Chart Source Data - Category (X) ...

Note:
*Your range selection, being made on the worksheet, is shown in the collapsed **Chart Wizard** dialog box.*

Note:
*Both the **Category (X)** range and **Value (Y)** range of the chart's data source must be of the same size.*

⑤	Select the **Category (X) Axis** Data Source (A22:A28): ◆ Click cell **A22**, ◆ Hold down [SHIFT] and click cell **A28**.	Chart Wizard - Step 2 of 4 - Chart Source Da =CarryHistogram!$A$22:$A$28 |	Bin	Frequency	Frequency
21			
22	251	1	1.4%
23	253	3	4.2%
24	255	15	20.8%
25	257	22	30.6%
26	259	17	23.6%
27	261	12	16.7%
28		2	2.8%	
⑥	Click the **Expand Dialog** button.	Source Data - Category (X) axis labels: =CarryHistogram!$A$22:$A$28
⑦	Back in the **Data Source** dialog box, click the **Next >** button.	Next >
⑧	The remaining steps should be familiar. Try to get your chart to look like the one shown in Figure 5.8.	Procedures 4.1, 4.2, ...

It is a good time to save your work.

Save

Compared to the default frequency distribution shown in Figure 5.3, this histogram appears to be smoother and more readable. It also better displays the approximately *bell shaped* pattern of the ball distances represented by the variable *Carry*. More than **95%** of all distances fall between **251** and **261** yards. The remaining distances may happen to be *outliers*. The next chapter provides an in-depth analysis of outliers.

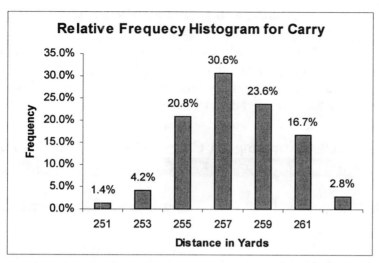

Figure 5.8 Relative Frequency Histogram for *Carry*

Exercise 3. Create a frequency histogram and a relative frequency histogram for the variable *TotDist* for all observations. How do these compare to the corresponding histograms for the variable *Carry*?

Section 5.5 Creating Histograms for Subsets of a Variable

The major objective in the analysis of the data set *Carry* is to determine which of the ball designs is "best". If we examine the histogram in Figure 5.8, we can not compare the variable *Carry* for the two different ball designs because we have displayed all **72** values of the variable *Carry* in one histogram. In order to compare the *Carry* of the ball for the two different ball designs (M1 and M2) we must somehow distinguish the data values in the graph. To this end, you will extract the M1 and M2 *Carry* values in separate ranges to another location. Next, employing the procedure demonstrated in the previous section, you will generate frequencies and histograms of *Carry* for models **M1** and **M2**.

Section 5.5.1 Extracting and Copying Data in Excel

To get started, you will extract the *Carry* values, corresponding to *Model* M1, and the *Carry* values, corresponding to *Model* M2, from the entire data set to a temporary location on the **Data** sheet. Then, you will copy the extracted data to **Sheet2**.

Procedure 5.9 Extracting M1 Carry Values from the Data Set

	Task Description	Mouse/Keyboard
❶	Switch to sheet **Data**.	
❷	On the right-hand side of the data set, create (type) two extraction **Filters** and output location for **M1** and **M2 Carry** values.	
❸	Select any cell of the data set range, for example **N2**.	
❹	Invoke the **Data | Filter | Advanced Filter** command.	
	*Excel opens the **Advanced Filter** dialog box. Note that based on the selected cell, Excel was able to **sense** the range of the entire data set ($A$1:$N$73).*	
❺	In the **Action** frame, select the **Copy to another location** option.	

Note:

*The filter for **Model** M1 contains the name of the variable (in **P1**) and its specific value (in **P2**). The cell **P3** determines **what** and **where** will be extracted (the output) .*

Note:
*Make sure the **Copy to another location** option button is turned on:*

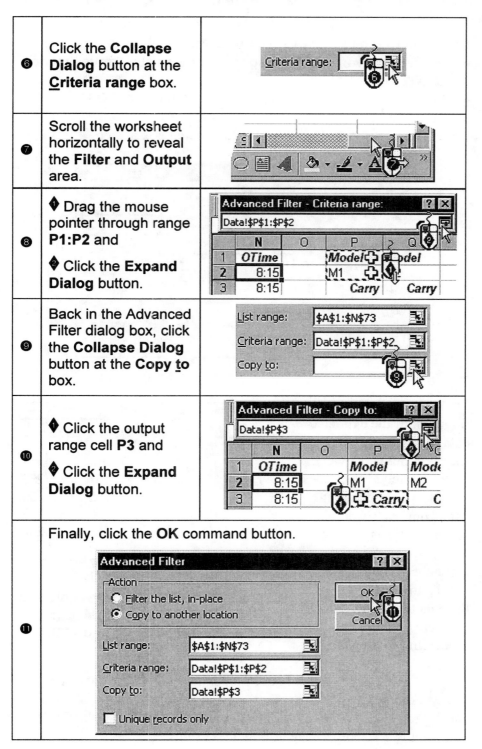

⑥	Click the **Collapse Dialog** button at the **Criteria range** box.
⑦	Scroll the worksheet horizontally to reveal the **Filter** and **Output** area.
⑧	◆ Drag the mouse pointer through range **P1:P2** and ◆ Click the **Expand Dialog** button.
⑨	Back in the Advanced Filter dialog box, click the **Collapse Dialog** button at the **Copy to** box.
⑩	◆ Click the output range cell **P3** and ◆ Click the **Expand Dialog** button.
⑪	Finally, click the **OK** command button.

Note:
In step ⑥, you may also type the range **P1:P2** *right in the* **Criteria range** *box. If you do, you will then be able to skip steps ⑦ and ⑧.*

Note:
Again, in step ⑨, if you type **P3** *right in the* **Copy to** *box, you will be able to skip step ⑩.*

Following the steps ❹ - ⓫ of the above procedure, extract the M2 *Carry* values. You will only need to make a few changes:

- In operation ◆ of step ❽: Drag the mouse pointer through range **Q1:Q2**.
- In operation ◆ of step ❿: Click the output range cell **Q3**.

Here is the state of the **Advanced Filter** dialog box at step ⓫.

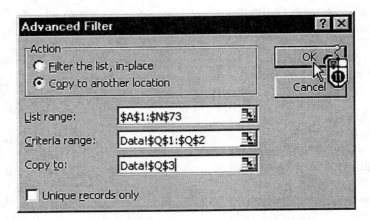

Figure 5.9 shows a fragment of the extracted data. Each subset of *Carry*, contains **36** values. Unfortunately, Excel does not allow for extracting data from one sheet to another. This is why you had to do it right on the **Data** sheet. Your next task will be to copy or move the data to **Sheet2**, which is where you will continue processing the subsets.

	N	O	P	Q
1	*OTime*		*Model*	*Model*
2	8:15		M1	M2
3	8:15		*Carry*	*Carry*
4	8:15		257	256
5	8:15		255	255
6	8:15		256	258
7	8:15		255	257
38	8:..			
39	8:45		257	255
40	8:45			

◀◀ ◀ ▶ ▶▶ \ CarryHistogram \ **Data** \ Sheet2 /

Figure 5.9 Fragment of the Carry subsets (for balls M1 and M2).

At this point, you should be an expert in copying data from one workbook location to another. Just in case, you may wish to check Procedure 4.10, which shows how to do a similar task. Or, if you want to try another way, follow these steps:

Procedure 5.10   Copying Data Between Sheets Using Keyboard Commands

Task Description	Mouse/Keyboard
❶ Go to cell **P3**: ◆ Press ⬚F5⬚ ◆ In the **Reference** box, type **P3** and press ⬚ENTER⬚.	⬚F5⬚ **Go To** ？✕ Go to: Ball Bin Carry Criteria Dd Dw Extract OK Cancel Special... Reference: P3 ❷ ⬚Enter⬚
❷ Select the two subsets: Hold down ⬚SHIFT⬚ and press ⬚END⬚, ⬚↓⬚, ⬚→⬚.	❷ ⬚Shift⬚ ⬚End⬚ ⬚↓⬚ ⬚→⬚
❸ Copy the selection to **Clipboard**: Hold down ⬚CTRL⬚ and press ⬚C⬚.	❸ ⬚Ctrl⬚ ⬚C⬚
❹ Switch to **Sheet2**: Hold down ⬚CTRL⬚ and press ⬚PAGE DOWN⬚	❹ ⬚Ctrl⬚ ⬚Page Down⬚
❺ Paste the contents of **Clipboard**: Hold down ⬚CTRL⬚ and press ⬚V⬚.	❺ ⬚Ctrl⬚ ⬚V⬚
❻ Clear the selection and stay in **A1**: Hold down ⬚CTRL⬚ and press ⬚HOME⬚.	❻ ⬚Ctrl⬚ ⬚Home⬚

Note:
You can move around between neighboring sheets by pressing ⬚CTRL⬚+⬚Page Dn⬚, ⬚CTRL⬚+⬚Page Up⬚.

Right away change the titles of the subsets from *Carry*, *Carry* to *CarryM1* and *CarryM2* and rename the sheet **Tab** as **CarryM1M2**. Now, enter new titles next to the subsets, as shown in Figure 5.10.

Hint:
To rename a sheet **Tab**, *double-click the* **Tab** *and enter a new name.*

	A	B	C	D	E
1	*CarryM1*	*CarryM2*	*BinM1M2*	*FrqCarryM1*	*FrqCarryM2*
2	257	256			
3	255	255			
4	256	258			
5	255	257			
6	255	257			
7	256	257			
8	255	258			
9	258	258			

◄ ◄ ► ►│ ╱ Data ╲ CarryM1M2 ╱ Sheet3 ╱ Sheet4 ╱ RangeDefDlg ╱│ ◄│

**Figure 5.10   Fragment of the M1 and M2 Carry
Subsets Brought to Sheet CarryM1M2**

You may wish to explore some or all of the following editing
tricks:

Procedure 5.11 A Few Text Editing Tricks in Excel

	Task Description	Mouse/Keyboard
❶	To change **Carry** to **CarryM1** in **A1**.	Select cell **A1**, press ⌨F2, type **M1** and press ⌨ENTER.
❷	To change **Carry** to **CarryM2** in **B1**.	Drag the **Fill Handle** of cell **A1** and drop it in **B1**.
❸	To apply the same format in cell **C1** as in **A1**: ◆ Type text BinM1M2 into cell **C1**, ❷ Click cell **A1**, ❸ Click the **Format Painter** button, and ◆ Click cell **C1**.	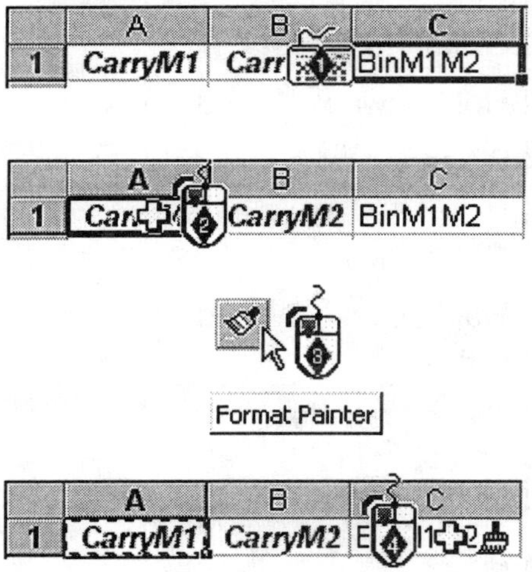

❹	To get text *FrqCarryM1* in **D1**: ◆ Press CTRL+HOME, ❷ Press CTRL+C for **Edit	Copy**, ❸ Press →,→,→ to move to **D1**, ❹ Press ENTER or CTRL+V for **Edit	Paste**, ❺ Press F2, to switch to the **Edit** mode, ❻ Press HOME to move the cursor in front of the first character, ❼ Type Frq, and press ENTER.	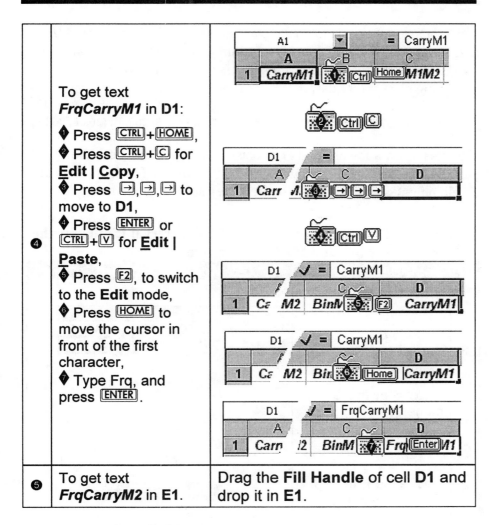
❺	To get text *FrqCarryM2* in **E1**.	Drag the **Fill Handle** of cell **D1** and drop it in **E1**.

It is a good time to save your work.

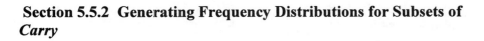

You should now have everything ready for generating the frequency distributions of the *CarryM1* and *CarryM2* subsets.

Section 5.5.2 Generating Frequency Distributions for Subsets of *Carry*

Using the procedure we learned in previous sections, you can now create two frequency histograms; one for the subset of the *Carry* values associated with model M1 (occupying range **A2:A37**) and a second one for the subset of the *Carry* values associated with model M2 (stored in range **B2:B37**).

Note:
In many statistical applications, a frequency histogram is used as a model of the probability distribution, where each interval is represented by its midpoint. In such cases, the external class intervals should be finite.

Our first task is to create class intervals (bins) for *CarryM1* and *CarryM2*. We will use a fewer number of class intervals and the two external intervals will be closed this time. Let us try the following class intervals:

(250.0, 252.5], (252.5, 255.0], (255.0, 257.5], (257.5, 260.0], (260.0, 262.5]

The reason for a fewer number of class intervals is that both the subsets (*CarryM1* and *CarryM2*) are smaller in size (each consists of **36** elements) compared to the entire *Carry* set (**72** elements).

Procedure 5.12   Entering a Series of Class Interval Limits for the M1 and M2 Carry Subsets

	Task Description	Mouse/Keyboard
❶	Enter the left limit of the first class interval: ◆ Go to **C2**, ◆ Type **250** and press CTRL + ENTER . ***Note***: Stay in the same cell.	
❸	Invoke the **Edit	Fill	Series** command.	
	Excel opens the **Series** *dialog box.*	
❹	In the **Step value** box, type the width of the class intervals, **2.5**, and press TAB .	Step value: 2.5
❺	In the **Stop value** box, type the right limit of the last class intervals, **262.5** (do **not** press ENTER).	Stop value: 262.5

⑥	Select the **Series in** **C**olumns option.	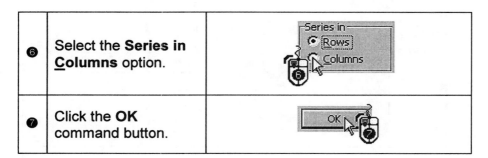
⑦	Click the **OK** command button.	

Note:
Make sure that the **C**olumns *option button is turned on:*

You will now define the frequency distribution for *CarryM1* using the **Frequency** function with a relative reference to the range of the *CarryM1* values (**A2:A37**) and with an absolute reference to the class interval (**M1M2Bin**) range (**$C$2:$C$7**). Such a setup will enable you to generate the frequency distribution of *CarryM2* by copying the **Frequency** function from **D2:D7** to **E3:E7**.

Note:
In Excel formulas, a $ sign used in front of a reference (column and/or row) makes sure that the reference will change when the formula is copied to another location.

Procedure 5.13 Generating Frequency Distributions for *CarryM1* **and *CarryM2***

Task Description	Mouse/Keyboard
① Select the *FrqCarryM1* range (**D2:D7**).	
② ◆ Type =**Frequency(** and ◆ Click cell **A2**.	
③ Use the vertical scroll bar to reveal the last row (**37**) of the input data.	

Note:
Excel encloses an array formula between curly braces, {}. No part of the array can be deleted. To remove the entire array, select its range and than press ⌦.

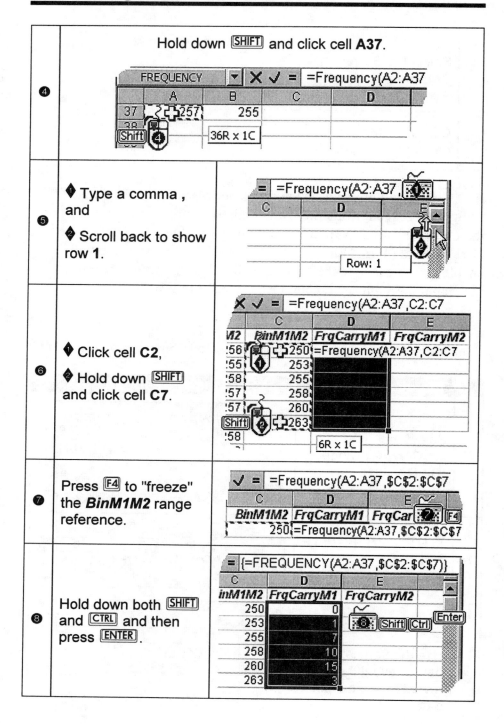

④ Hold down ⇧SHIFT and click cell **A37**.

⑤ ◆ Type a comma ,
and

◆ Scroll back to show row **1**.

⑥ ◆ Click cell **C2**,

◆ Hold down ⇧SHIFT and click cell **C7**.

⑦ Press F4 to "freeze" the ***BinM1M2*** range reference.

⑧ Hold down both ⇧SHIFT and CTRL and then press ENTER.

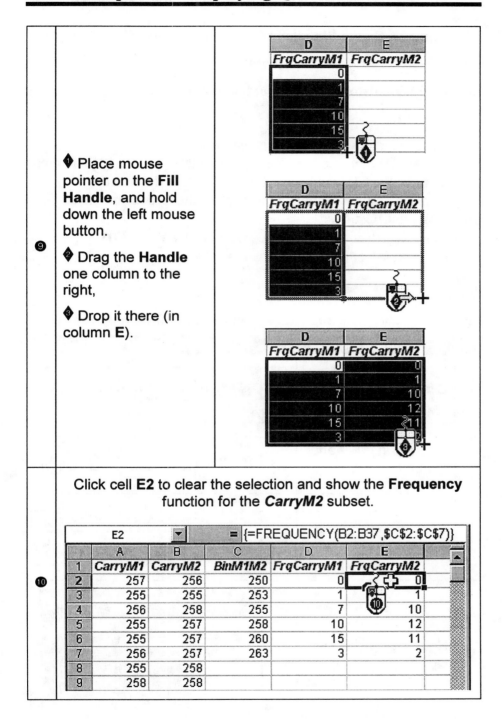

♦ Place mouse pointer on the **Fill Handle**, and hold down the left mouse button.

♦ Drag the **Handle** one column to the right,

♦ Drop it there (in column **E**).

Click cell **E2** to clear the selection and show the **Frequency** function for the *CarryM2* subset.

E2			= {=FREQUENCY(B2:B37,$C$2:$C$7)}

	A	B	C	D	E
1	CarryM1	CarryM2	BinM1M2	FrqCarryM1	FrqCarryM2
2	257	256	250	0	0
3	255	255	253	1	1
4	256	258	255	7	10
5	255	257	258	10	12
6	255	257	260	15	11
7	256	257	263	3	2
8	255	258			
9	258	258			

As you can see the second **Frequency** function shown on the **Formula** bar is correct. It takes the **Data** input from **CarryM2** in the

B2:B37 range and the **Bin** input from the same range (**C2:C7**) as the first function.

Section 5.5.3   Histograms for Comparing the Frequency Distributions

Note:
The **Frequency** *functions have determined that there are no values below* **250**. *This is why the two zero cells* (**D2:E2**) *are ignored in definition of the chart's input.*

You can now create two histograms at one time by selecting the title (**D1:E1**) range and the **Value (Y) Axis** range (**D3:E7**) and then running the **Chart Wizard** according to the steps described by Procedure 5.8. When following the steps of that procedure, make the following modifications:

- In step ❶, select two non-adjacent ranges **D1:E1** and **D3:E7**,
- In step ❺, select range **C3:C7**.

Hint:

Procedure 5.14 Selecting No-adjacent Ranges for Chart Input

	Task Description	Mouse/Keyboard
❶	Click **D1**.	
❷	Hold down ⬛SHIFT⬛ and click **E1**.	
❸	Hold down ⬛CTRL⬛ and click **D3**.	
❹	Hold down ⬛SHIFT⬛ and click **E7**.	

Figure 5.11 shows the resulting histograms.

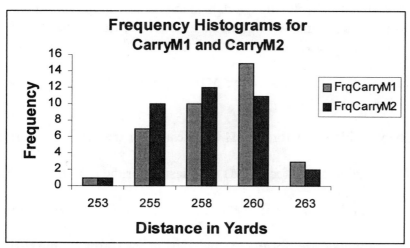

Figure 5.11   Frequency Histograms for the M1 and M2 Subsets of *Carry*.

Exercise 4. Compare the histograms shown in Figure 5.11. What can you conclude about the behavior of the variable *Carry* for the two different ball designs?

When you are comparing histograms you need to be sure that you are making a fair comparison. That is you would really like to guarantee that the class intervals shown along the **X**-axis are the same in both graphs. In this case, we intentionally created common class intervals. In order to make sure that your histograms are comparable when using the **Tools | Data Analysis | Histogram** command, you should define the same class interval range in the **Bin Range** input box of the **Histogram** dialog box. If the **Bin Range** box is left undefined, Excel creates its own class intervals. For even slightly different samples the default class intervals will most likely not be identical.

Section 5.6 Investigative Exercises

In the following exercises you are asked to create a variety of different graphs which will display various aspects of the golf ball data file. The graphs are then to be used to draw conclusions about the data file and make recommendations to management.

DO NOT treat the individual exercises or even the individual components of an exercise as isolated graphs. Remember the reason you are visually displaying the data is to enable you to "see" the information contained within the data. Look for trends, patterns, similarities, and differences between the graphs.

1. In a workbook to be saved as **M1Carry.xls**, construct a histogram for:

a) The variable *Carry* for the **M1** ball design for the 8:15 time period.

b) The variable *Carry* for the **M1** ball design for the 8:45 time period.

c) The variable *Carry* for the **M1** ball design for the 9:15 time period.

d) The variable *Carry* for the **M1** ball design for all three time periods combined.

e) Does the variable *Carry* exhibit generally the same behavior in each time period?

f) What, if any, differences did you see?

---

2. Generate four histograms similar to those constructed in exercise 1 using the **M2** ball design. Save the worksheet as **M2Carry.xls**.

a) Graph **M2Carry0815**:

b) Graph **M2Carry0845**:

c) Graph **M2Carry0910**:

d) Graph **M2Carry**:

e) Does the variable *Carry* exhibit generally the same behavior in each time period?

f) What, if any, differences did you see?

3. Compare the graphs from exercise 1 with those generated in exercise 2. What can you conclude about the difference, if any, between the two ball designs with regard to how far the ball carries?

---

4. Generate histograms to display the total distance the **M1** ball traveled during each of the three time periods individually and combined. Save the worksheet as **TotDist1.XLS**.

a) Graph **M1TotDist0815**:

b) Graph **M1TotDist0845**:

c) Graph **M1TotDist0910**:

d) Graph **M1TotDist**:

e) Does the variable *TotDist* exhibit generally the same behavior in each time period?

f) What, if any, differences did you see?

---

5. Generate four histograms similar to those constructed in exercise 4. Use the **M2** ball design. Save the worksheet as **TotDist2.XLS**.

a) Graph **M2TotDist0815**:

b) Graph **M2TotDist0845**:

c) Graph **M2TotDist0910**:

d) Graph **M2TotDist**:

e) Does the variable *TotDist* exhibit generally the same behavior in each time period?

f) What, if any, differences did you see?

_____

6. Compare the graphs generated in exercise 4 with those generated in exercise 5. What, if any, differences are there in the total distance that the ball flies for the two ball designs?

_____

7. Experiment with different class intervals. For instance, try creating histograms of the variable *TotDist* for the **M1** golf ball with the interval width of **1** yard, then using **5** yards, and then **10** yards. Compare these 3 graphs. Save the worksheet as **TotDist7.XLS**.

a) Which graph gives management the most informative view of the data? Why did you select this one?

b) What happens to the histogram with the smallest class interval width (resulting in too many class intervals)?

c) What happens to the graph with the largest class interval width (resulting in too few class intervals)?

8. Construct a histogram of the variable *Ball*. Save the workbook as **Ball.XLS**.

a) What does it look like?

b) Why should you have expected it to look like this?

c) Why is it a useless graph?

---

9. Change the values of *Carry* and *TotDist* for *Ball* **1** to **357** and **370** respectively.

a) Create a histogram of the variable *Carry* for the **M1** ball design for the time period 8:15.

b) Compare this to the graph saved in **M1Carry.XLS**. What has happened to the histogram as a result of this unusual data point?

---

10. Visually examine the variable *Temp* in the spreadsheet.

a) What do you notice?

b) Construct a histogram of all **72** temperature values.

c) Does the histogram show what you noticed about the temperatures?

d) Create a different way of displaying the temperatures which will in fact display what you noticed when you visually examined the temperature values.

11. Of what use is it to have the variable *ODate* when they are all the same?

_____

12. Look at the variables *Carry* and *TotDist* for ball numbers **64** and **66**.

a) What do you notice?

b) Assuming you have correctly controlled for all the factors, what is causing the variation in the total distance even though the ball carried the same yardage in both cases?

_____

13. Construct whatever histograms you need to examine whether any of the controlled factors, internal of external, were really not held constant during the test.

_____

14. On the basis of your analysis of the histograms you created, which of these two ball designs would you recommend to management. Support your answer with the appropriate graphs that illustrate your points.

Chapter 6   "Golf Ball Design"

Numerical Descriptors

Section 6.1 Overview

In the last two chapters you have studied graphical methods for displaying data. Although these methods provide a visual picture of the data, they do not provide any numerical, summary information about the data. This chapter will focus on the most commonly used summary statistics. These are also called numerical descriptors or descriptive statistics. You will see how to use **MEASURES OF THE MIDDLE** and **MEASURES OF DISPERSION** to describe a large data set.

Statistical Objectives: After reading this chapter and doing the exercises a student will:
- Know how to use the mean as a measure of the middle and understand its limitations.
- Know how the median differs from the mean and in what situations it is a better measure of the middle.
- Know that by comparing the mean and the median you can tell the general shape of the distribution.
- Know the meaning of the variance and the standard deviation.
- Know that the **z**-score tells you how many standard deviations the data point is from the mean.
- Know how to use the empirical rule.

Section 6.2 Problem Statement

We return to the data set from Chapter 5 concerning two different golf ball designs. Recollect that we introduced a manufacturer interested in product design and quality improvement who was collecting a lot of data on the performance of its golf balls. The actual variables of interest to which the firm was directing its research were flight and total distance the ball carries. W. Edwards Deming, a founding statistician of the modern quality assurance movement, has commented that raw data itself is meaningless without the analyst having strong skills in understanding its most difficult feature, random variability. Developing the skills of interpretation and prediction with sample data, Deming would say, are the keys to understanding and controlling product variability and to managing consistent product improvement. As you develop numerical descriptors to summarize the golf ball data file, keep Deming's insights in mind.

Section 6.3 Generating Descriptive Statistics for an entire column of data

Section 6.3.1 Excel command to calculate summary statistics

*Click the **Open** button or use the **File** | **Open** command (ALT + F, O) to open the file.*

Open the data set named **Ch06Dat.xls** from your data disk. Note that Procedure 3.1, on page 33, contains detail instruction about how to open an Excel workbook.

To simplify referencing the variables (samples) in the data set, name all the variables using the names stored in the top row. Just follow the steps of Procedure 5.1 (page 100). You may wish to inspect the names by pulling down their list from the formula bar. Click the drop-down arrow located on the right-hand side of the name box, and then click one of the names. Excel will select the range associated with the selected name. The following procedure shows an example for selecting the range named as **Model**.

Procedure 6.1 Selecting a Named Range from the Name Box

Task Description	Mouse/Keyboard
❶ ◆ Click the **Drop-down** arrow at the **Name** box, located on the left-hand side of the **Formula** bar. ◆ Choose the name from the list.	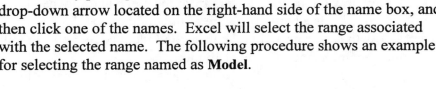

In Excel, the statistical summary measures can be calculated using Excel's intrinsic functions or via the **Tools** | **Data Analysis** | **Descriptive Statistics** command. You will first explore basic summary measures for the *Carry* variable using the **Data Analysis** tools.

**Procedure 6.2   Generating Basic Summary Measures of Carry
Using the Data Analysis Tools**

	Task Description	Mouse/Keyboard
❶	Invoke the ◆ **Tools** ◆ **Data Analysis** command.	
	*Excel opens the **Data Analysis** dialog box.*	
❷	Choose option **Descriptive Statistics**.	
	*Excel opens the **Descriptive Statistics** dialog box.*	
❸	In the **Input Range** box, type **Carry** (**do not** press [ENTER]).	
❹	In the **Output options** frame do this: ◆ Click the **New** **Worksheet Ply** box, ◆ Type **SummaryOfCarry** ◆ Click the **Summary** **statistics** check box.	

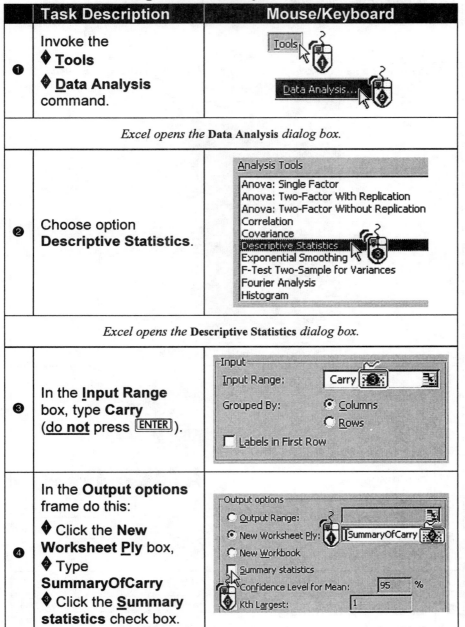

Note:
🖱 *Double-clicking an
option in a dialog box
is often equivalent to
🖱 clicking both the
option and the **OK**
command button.*

Note:
*Make sure the
Summary statistics box
is checked:*

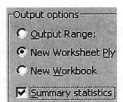

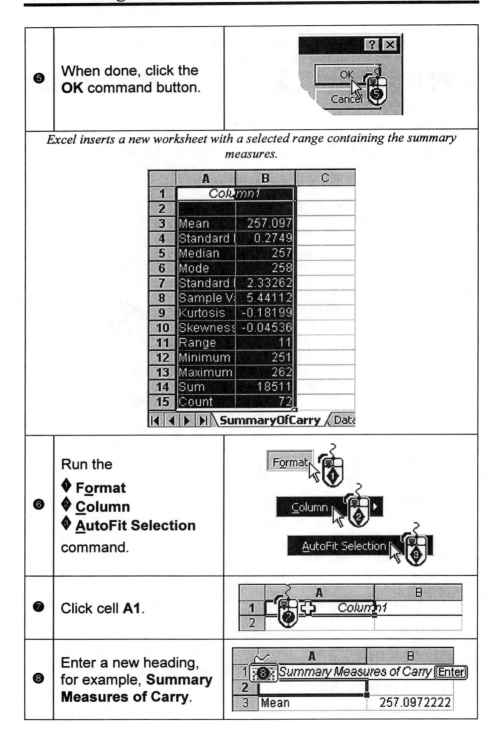

| ❺ | When done, click the **OK** command button. | |

Excel inserts a new worksheet with a selected range containing the summary measures.

	A	B	C
1	*Column1*		
2			
3	Mean	257.097	
4	Standard [	0.2749	
5	Median	257	
6	Mode	258	
7	Standard [	2.33262	
8	Sample V	5.44112	
9	Kurtosis	-0.18199	
10	Skewness	-0.04536	
11	Range	11	
12	Minimum	251	
13	Maximum	262	
14	Sum	18511	
15	Count	72	

SummaryOfCarry / Data

| ❻ | Run the
◆ **Format**
◆ **Column**
◆ **AutoFit Selection**
command. | |

| ❼ | Click cell **A1**. | |

| ❽ | Enter a new heading, for example, **Summary Measures of Carry**. | |

Figure 6.1 shows the final view of the summary measures. Excel has generated a static report, including the basic statistical

	A	B	C
1	Summary Measures of Carry		
2			
3	Mean	257.0972222	
4	Standard Error	0.274901976	
5	Median	257	
6	Mode	258	
7	Standard Deviation	2.332620615	
8	Sample Variance	5.441118936	
9	Kurtosis	-0.181989376	
10	Skewness	-0.045362901	
11	Range	11	
12	Minimum	251	
13	Maximum	262	
14	Sum	18511	
15	Count	72	

|◄ ◄ ► ►| **SummaryOfCarry** / Data / BoxPlot / B

Figure 6.1 Descriptive Statistics Outcomes for Carry

summary measures. If the input data set gets changed, the summary measures will have to be regenerated by running the **Tools | Data Analysis | Descriptive Statistics** command again.

Section 6.3.2 "Measures of the Middle"

Excel generates three commonly used measures of the middle of a set of data. These are the *mean*, *median*, and *mode*. The mean of the variable *Carry* is approximately equal to **257.1**, the median is equal to **257**, and the mode is equal to **258**. Note that the mode is usually not a good measure of the middle, when calculated directly from a sample of non-discrete numbers. A better representation can be derived from the frequency distribution.

Many times when data are analyzed, the mean and the median are calculated and quoted but never really examined. Take a few minutes to *see* what information is actually contained in these two numerical descriptors.

Recall that the mean is found by adding up all the data values and dividing by the number of observations. Also remember that the median is found by sorting the data from smallest to largest and selecting the observation which falls right in the middle of the ranked

The Mean can also be calculated using the Average() function, here:
    =Average(Carry) .

The Median can also be calculated using the Median() function, here:
    =Median(Carry) .

The Mode can also be calculated using the Mode() function, here:
    =Mode(Carry) .

data. Think about what will happen to the mean and the median if there are one or two really large numbers in the data set.

In order to see the effect of so called *outliers* you need to look at what portion of the formulas will be impacted. In the case of the median, a large value does not have any impact because the median is simply the middle score. It does not matter how big or how small the extreme values are, the median will still be the same.

In the case of the mean, a large observation will cause the mean to be higher because the sum of the data values will be higher. The **Descriptive Statistics** report also provides the value for the sum. As you can see in Figure 6.1, Sum = **18511**. When this is divided by the number of observations, which is shown as Count (right below Sum), you get the mean value of **257.1**.

> **Exercise 1:** On the **Data** sheet, change the value in the first row of the variable *Carry* to **387**. Re-issue the **Tools | Data Analysis | Descriptive Statistics** command. What happens to the sum? What happens to the mean?

Hint: Very carefully apply the following steps:

Procedure 6.3   Re-generating Basic Summary Measures of Carry Using the Data Analysis Tools

Task Description	Mouse/Keyboard
❶ ◆ Click the **Drop-down** arrow at the **Name** box. ◆ Select Name **Carry**.	A2 / Ball / Carry / Dd / Dw / Head / Model / ODate
*Excel jumps to the **Data** sheet, selects the range named as Carry, and activates the first cell (**K2**) in the range.*	
❷ Type **387** in **K2** and press ENTER.	J / K / L — 1 Temp / Carry / TotDist; 2 / 387 / 270; 3 77 / 255 / 267

Sidebar:

The Sum *can also be calculated using the* **Sum()** *function, here:* =**Sum(Carry)** .

The Count *(**Sample Size**) can also be calculated using the* **Count()** *function, here:* =**Count(Carry)** .

Note: The **Descriptive Statistics** *dialog box has not changed since it was first introduced in with Excel 5.0. It is not an intrinsic part of the Excel command system. It is an Add-In component, which is why it exhibits a kind of a strange behavior, especially when used for the second time during the same Excel session.*

❸	Switch back to sheet **SummaryOfCarry**.	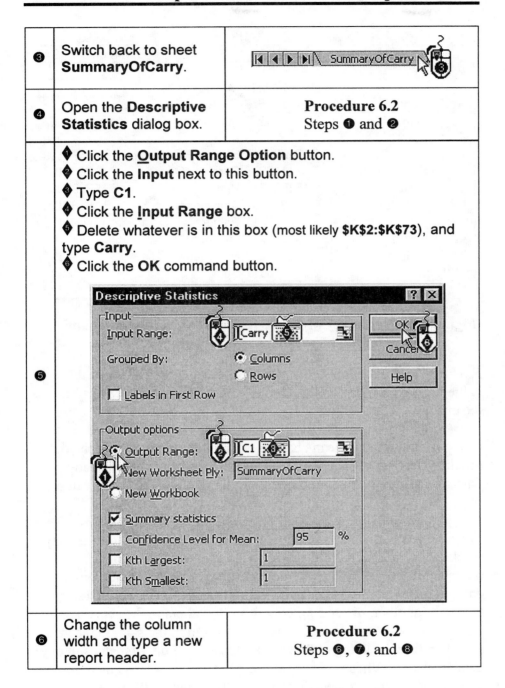
❹	Open the **Descriptive Statistics** dialog box.	**Procedure 6.2** Steps ❶ and ❷

| ❺ | ◆ Click the **Output Range Option** button.
◆ Click the **Input** next to this button.
◆ Type **C1**.
◆ Click the **Input Range** box.
◆ Delete whatever is in this box (most likely **$K$2:$K$73**), and type **Carry**.
◆ Click the **OK** command button. | |

| ❻ | Change the column width and type a new report header. | **Procedure 6.2** Steps ❻, ❼, and ❽ |

Note:
*Make sure the **Output Range Option** button and **S**ummary statistics **Check** box are set on.*

Figure 6.2 shows the new summary measures report.

What happened to the mean of *Carry*?

Note:
*When done, change the value in the first row of the variable **Carry** back to **257**.*

	A	B	C	D
1	*Summary Measures of Carry*		*Summary Measures of Carry2*	
2				
3	Mean	257.0972222	Mean	258.9027778
4	Standard Error	0.274901976	Standard Error	1.825008813
5	Median	257	Median	257
6	Mode	258	Mode	258
7	Standard Deviation	2.332620615	Standard Deviation	15.48571329
8	Sample Variance	5.441118936	Sample Variance	239.8073161
9	Kurtosis	-0.181989376	Kurtosis	68.63299093
10	Skewness	-0.045362901	Skewness	8.18982403
11	Range	11	Range	136
12	Minimum	251	Minimum	251
13	Maximum	262	Maximum	387
14	Sum	18511	Sum	18641
15	Count	72	Count	72

| ◄ ◄ ► ►| \ **SummaryOfCarry** / Data / BoxPlot / | ◄ | | ► |

Figure 6.2   Summary Measure Reports for the Original and Modified Variable Carry

Exercise 2: What happened to the median when you changed the value in row **1** of the variable *Carry* ? Recall that it was **257** prior to the change. What is it now?

Although the mean changes quite a bit, the middle of the data (median) has not changed at all. Thus, the mean can be misleading if what you are trying to communicate is the middle of the data values.

Unusually large or small observations are called *outliers*. They lie very far from the middle of the distribution. An outlier may result from transposing digits when recording or observing an observation. Even if there are no recording or observational errors, a data set may contain one or more valid measurements which, for one reason or another, differ markedly from the others in the set. As you have seen, these outliers can cause a distortion in the sample mean, $\bar{x}$. Therefore, isolating outliers should be one of the first steps in any data analysis.   When there are outliers in the data set, the median gives a more accurate description of the "middle" of the data.

The next question you should ask is "How can I tell if an observation is an outlier?" The measures of dispersion and box plots discussed in the next two sections will help you answer this question.

However, before leaving the measures of the middle you should notice that prior to changing the value in row **1** of the variable *Carry*, the mean and the median were almost equal in value.

Exercise 3: Examine the descriptive statistics of *Carry* shown in Figure 6.2. Compare the mean and median values.

 In general, they will not be equal but you should ***always*** compare their values. By comparing them you can get an initial clue as to whether or not there are any outliers. Secondly, by comparing the mean and the median you can get a sense of the shape of the data. Often times, the mean and the median are quoted but you do not have access to the data directly and a graph may not be provided. In this case, a simple comparison of these two values will give you a general idea about the graph.

 When the mean and the median are close in value, then the underlying distribution is symmetric and often has a bell shape. This results in a skewness value close to a zero. The *Carry* variable exhibits a negligible skewness, the measure of Skewness = -.04536 (as seen in Figure 6.2). If the mean is much smaller than the median then the shape of the graph is skewed to the left. This means that there are a few extremely low values which are dragging the mean down while the median is unaffected by these low values. The measure of Skewness would then be significantly negative. If the mean is much larger than the median then the graph is skewed to the right. This means there are a few large values, which are pulling the mean up while the median is a much smaller value. In such cases, the measure of Skewness would be significantly positive.

The Skewness can also be calculated using the **Skew()** *function, here:*
  =Skew(Carry)

Exercise 4: Inspect the frequency histogram of the variable *Carry*, shown in Figure 5.8 (page 117). Does the graph confirm what you expected?

 In order to decide whether or not a particular statistical technique can be used, it is often important to know whether the distribution has a symmetric shape. This matter will be further discussed in Chapters 10-14.

Section 6.3.3 "Measures of Variability"

The
Standard Deviation
can also be calculated
using the **StDev()**
function, here:
        = StDev(Carry)
Note: **StDev()** *is a*
sample standard
deviation. To compute
the population
standard deviation use
the **StDevP()** *function.*

In the output shown in Figure 6.2, you can also find measures of variability. The most important measure of variability is the standard deviation. For the *Carry* variable it is shown as **2.33**. Information is also provided on the minimum and maximum values of the variable. A quick subtraction of the minimum value (shown as **Min**) from the maximum value (shown as **Max**) gives you the **Range** of the data which is another frequently quoted measure of variability. In this case, the range of the data is **262-251=11**.

Exercise 5: Be sure you can find the standard deviation, minimum, and maximum in the output.

As was the case with the measures of the "middle", the measures of dispersion are often simply calculated and then not used because people do not really understand what information they contain. Take a closer look at each of these measures to get a better idea of what they really tell you about the data. Remember your job is to extract as much information from the data in order to be able to make informed decisions.

The Minimum
can also be calculated
using the **Min()**
function, here:
        =Min(Carry)

The Maximum
can also be calculated
using the **Max()**
function, here:
        =Max(Carry)

The range is the easiest measure of variability to calculate and is therefore often quoted. Extreme care should be exercised in drawing any conclusions about the data on the basis of the range. By its very definition, **Range = maximum score - minimum score**, it is highly sensitive to the extreme values in the data set. By changing either the largest or smallest value or both, the range is quickly and often radically changed even though the amount of true variation in the data has not changed that much.

Exercise 6: Look at the output generated in Exercise 1, when you changed one value from **257** to **387**. Compare the **Max, Min**, and the **Range** values.

You should notice that the range has changed rather dramatically from its original value of **11** although the only thing that has really changed is one data observation! This characteristic makes the range a poor measure of dispersion.

The sample variance is often quoted but is useful primarily as a stepping stone to the standard deviation. By looking at the formula for the sample variance:

$$s^2 = \sum_{i=1}^{n}(x_i - \bar{x})^2 / (n-1)$$

you can easily see that the unit of measure on the variance is whatever unit of measure you were using squared. It is not particularly intuitive or helpful to think about dispersion in terms of units such as feet squared. Thus, the sample variance is most helpful simply as a step in getting the standard deviation. The standard deviation, **s**, is simply the square root of the variance: $s = \sqrt{s^2}$.

*The Variance can also be calculated using the **Var()** function, here:*
*= **Var(Carry)***
*Note: **Var()** is a sample variance. To compute the population variance use the **VarP()** function.*

The standard deviation is the most useful of all the measures of dispersion. Think about the standard deviation as a yardstick to be used to measure any and all differences in the data set. For example, suppose the distance between a data value and the sample mean is **100** feet. You can not tell by simply looking at the value of **100** feet whether the data value is close to the mean or far away from the mean. This is because all differences are only meaningful when compared to the standard deviation. What you want to know is to how many standard deviations does **100** feet correspond. If the standard deviation were **50** then the difference of **100** would translate to **100/50** or **2** standard deviations. However, if the standard deviation were **20** then the difference of **100** would translate to **100/20** or **5** standard deviations.

Note:
*The **standard deviation** is the most useful measure of dispersion.*

What are these numbers **2** and **5** called? You know them as the *z-score* for the observation. Remember that the **z**-score is simply the difference between the observation (x) and the sample mean ($\bar{x}$) divided by the standard deviation (s):

$$\text{z-score} = \frac{x - \bar{x}}{s}$$

If you take a sample from a Normal population, approximately **68%** of all observations should fall within **1** standard deviation of the mean. This means that **68%** of all observations should have a **z**-score between -**1** and **1**. Moreover, about **95%** of the observations should have a **z**-score between -**2** and **2** . Finally, virtually all of the observations (more than **99%**) should fall within **3** standard deviations of the mean and thus have a z-score between -3 and 3. This implies that a z-score of 2 is reasonable but a z-score of 5 is clearly an outlier.

Exercise 7: For the variable *Carry* which you have been looking at, how many standard deviations away from the mean is the observation **255**? Based on this calculation determine if it is an outlier.

Exercise 8: Calculate the z-score for **Ball #15**. Verify that it falls within one standard deviation of the mean by noticing that **Ball #15** carried a distance of **258** which is between the limits:

Did you get **Mean + 1** standard deviation and **Mean - 1** standard deviation?

Thus a z-score between **-1** and **1** tells you that the observation falls within one standard deviation of the mean.

The Kurtosis
can also be calculated
using the **Kurt()**
function, here:
= **Kurt(Carry)**

The **Descriptive Statistics** output also includes *kurtosis*, which characterizes the relative flatness/peakedness of data compared to the normally distributed data. A positive value of kurtosis indicates a relatively peaked distribution. A negative value indicates a relatively flat distribution. For the "normal" data, the kurtosis value is equal to 3. Chapter 8 explores the normal distribution in more detail.

Section 6.4   Generating Descriptive Statistics for Grouped Data

In Section 6.3 you learned how to use Excel to find descriptive statistics for a complete column of data. It is often useful to be able to calculate descriptive statistics for subsets of a column of data. For example, with the variable *Carry* we would definitely want to look at the summary measures of Carry for the different model types. Thus we want the measures of the variable *Carry* for model **M1** and model **M2**. Excel does not provide a direct way to determine all the measures for subsets of a numeric variable. We ran into this issue in Chapter 5 when we wanted to create a histogram for *Carry* for each model. Recall that in that case we had to extract the *Carry* data separately for *Model* M1, and M2).

The following steps explain how to compute the summary measures of the variable *Carry* separately for model **M1** and model **M2**.

Procedure 6.4   Computing Summary Measures for M1 and M2 Subsets of Carry

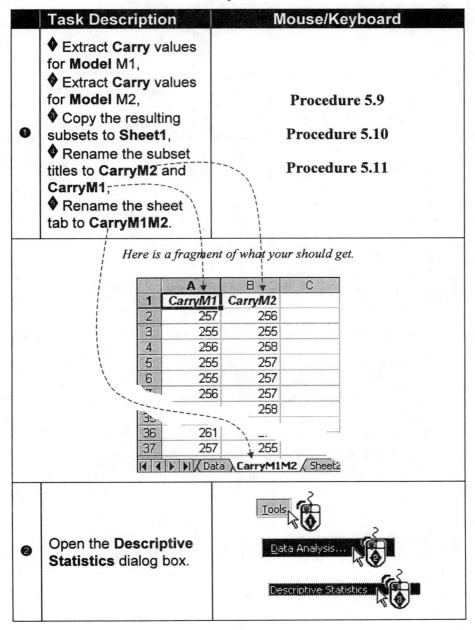

	Task Description	Mouse/Keyboard
❶	◆ Extract **Carry** values for **Model** M1, ❷ Extract **Carry** values for **Model** M2, ◆ Copy the resulting subsets to **Sheet1**, ◆ Rename the subset titles to **CarryM2** and **CarryM1**, ◆ Rename the sheet tab to **CarryM1M2**.	Procedure 5.9 Procedure 5.10 Procedure 5.11

Here is a fragment of what your should get.

	A	B	C
1	CarryM1	CarryM2	
2	257	256	
3	255	255	
4	256	258	
5	255	257	
6	255	257	
	256	257	
		258	
35			
36	261		
37	257	255	

◄ ◄ ► ►◄ \ Data \ **CarryM1M2** \ Sheet2

| ❷ | Open the **Descriptive Statistics** dialog box. | Tools
 Data Analysis...
 Descriptive Statistics |

Hint:
The **Tools | Data Analysis | Descriptive Statistics** *command can generate its output for a group of data sets, provided the sets are stored in adjacent columns.*
Note:
Another way for getting the CarryM1 and CarryM2 subsets would be to copy them from the Ch05Dat.xls workbook.

Hint:
If you open more than one workbook in Excel, you can switch between them via the **Window** *option*

Window	Help
New Window	
Arrange...	
Hide	
Unhide...	
Split	
Freeze Panes	
1 Ch05Dat.xls	
2 Ch06Dat.xls	

or by pressing CTRL + TAB.

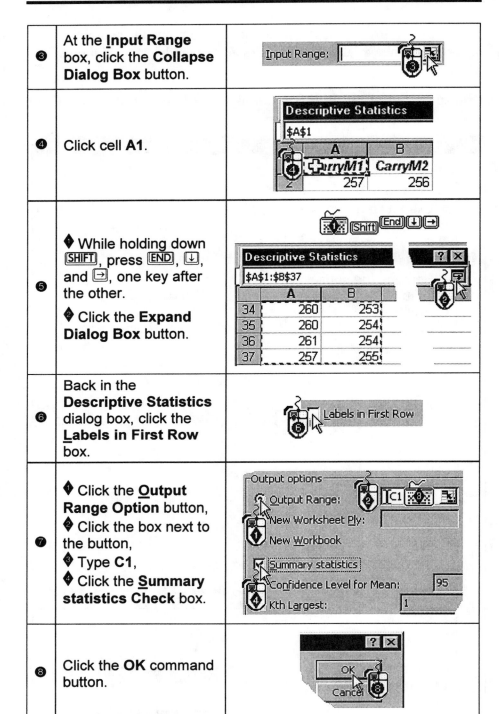

❸	At the **Input Range** box, click the **Collapse Dialog Box** button.	Input Range:
❹	Click cell **A1**.	**Descriptive Statistics** $A$1 ... A CarryM1 257 / B CarryM2 256
❺	◆ While holding down SHIFT, press END, ↓, and →, one key after the other. ◆ Click the **Expand Dialog Box** button.	SHIFT END ↓ → **Descriptive Statistics** $A$1:$B$37 ... 34 260 253 / 35 260 254 / 36 261 254 / 37 257 255
❻	Back in the **Descriptive Statistics** dialog box, click the **Labels in First Row** box.	Labels in First Row
❼	◆ Click the **Output Range Option** button, ◆ Click the box next to the button, ◆ Type **C1**, ◆ Click the **Summary statistics Check** box.	Output options ... Output Range: C1 ... New Worksheet Ply: ... New Workbook ... Summary statistics ... Confidence Level for Mean: 95 ... Kth Largest: 1
❽	Click the **OK** command button.	OK Cancel

Hint:
When the variable names (top labels) are included in the **Input Range**, *make sure that the* **Labels in First Row** *check box is set* **on**:

Note:
Make sure the **Summary statistics** *box is checked:*

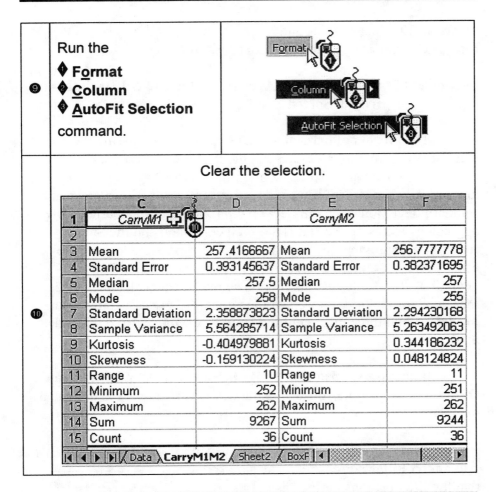

Run the

⑨ ♦ **F**ormat
    ♦ **C**olumn
    ♦ **A**utoFit Selection

command.

Clear the selection.

	C	D	E	F
1	*CarryM1*		*CarryM2*	
2				
3	Mean	257.4166667	Mean	256.7777778
4	Standard Error	0.393145637	Standard Error	0.382371695
5	Median	257.5	Median	257
6	Mode	258	Mode	255
7	Standard Deviation	2.358873823	Standard Deviation	2.294230168
8	Sample Variance	5.564285714	Sample Variance	5.263492063
9	Kurtosis	-0.404979881	Kurtosis	0.344186232
10	Skewness	-0.159130224	Skewness	0.048124824
11	Range	10	Range	11
12	Minimum	252	Minimum	251
13	Maximum	262	Maximum	262
14	Sum	9267	Sum	9244
15	Count	36	Count	36

Data \ **CarryM1M2** \ Sheet2 \ BoxF

Exercise 9: Are the measures significantly different? Compare the measures of shape (**Kurtosis** and **Skewness**) with the frequency histograms of *CarryM1* and *CarryM2*. Are the measures consistent with your visual perception?

It is a good time to save your workbook.

Save

We may want to further investigate the behavior of the average or mean *Carry* for different time periods within a particular model design. For example, we may want to look at only those values of *Carry* for the M1 ball design for the 8:15 time period. The steps for creating descriptive statistics for *CarryM1* at *OTime* = 8:15 would be very similar to Procedure 5.9 (page 118). In this case, you would need to involve the three sets: *Model*, *Carry*, and *OTime*, along with the criterion range based on both *Model* = M1 and *OTime* = 8:15.

Example:
Criteria
Range = **P1:Q2**
Output Range = **P3**

	P	Q
1	*Model*	*OTime*
2	M1	8:15
3	*Carry*	

Exercise 10: Generate the summary measures for the *Carry* values associated with *Model* = **M1** and time period *OTime* = **8:15**.

Section 6.4.5   Generating the Average and Standard Deviation from a Sorted List

Note:
Use the **Data** | **Subtotals** *command to calculate group averages, standard deviations, and/or other measures for a sorted (grouped) data set.*

The process of extracting data subsets and preparing them to generate the summary measures via the **Descriptive Statistics** command may be quite complex and time consuming. It could be dramatically simplified, if we reduced our interest to the average and standard deviation only.   In Section 4.8 (page 83), you learned how to aggregate data using the **Data** | **Subtotal** command and the **Sum** function. Employing this command, it is also possible to generate other aggregates, in particular—the average and the standard deviation. The following instruction shows how to calculate these measures for the three subsets of the variable *Carry* associated with *Model* M1 and three different time periods, *OTime* 8:15, 8:45, and 9:10.

If necessary, open the **Ch06Dat.xls** workbook and switch to the **Data** sheet. This time you will extract all necessary data in a manual manner.

However, should you decide to use the **Data** | **Filter** | **Advanced Filter** command, make sure that the filter setup is defined as in this example:

Criteria Range = P1:P2
Output Range = P3:Q3.

	P	Q
1	*Model*	
2	M1	
3	*Carry*	*OTime*
4		

The following procedure shows how to do it differently.

Procedure 6.5   Extracting Subsets of the Data Set Using the Copy, Sort, and Delete Commands

	Task Description	Mouse/Keyboard
❶	Unless your Excel window is already maximized, click the (Parent) **Maximize** button.	

②	On the **Data** sheet: ◆ Click the **Header** of column **B**, ❷ Hold down CTRL and click the **Header** of column **K**, ❸ Hold down CTRL and click the **Header** of column **N**, ❹ Click the **Copy** button, and ❺ Switch to **Sheet2**.	Copy 
❸	Click the **Paste** button.	 Paste
④	◆ Click any cell in column **A**, and ❷ On the **Standard Toolbar**, click the **Sort Ascending** button.	Sort Ascending 
④	Select rows **38 - 73**: ◆ Drag the vertical **Scroll** button down to row **38**, ❷ Click the row **38 Number**, ❸ Point to the **Scroll Down** arrow, press and keep holding down the left mouse button until row **73** is revealed, ❹ Hold down SHIFT and click the row **73 Number**.	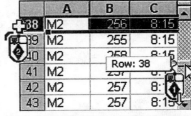

⑤	To delete the unwanted rows: ♦ Point anywhere to the selected row **Numbers** and click the <u>right</u> mouse button, ♦ From the **Shortcut** menu, select option **Delete**.	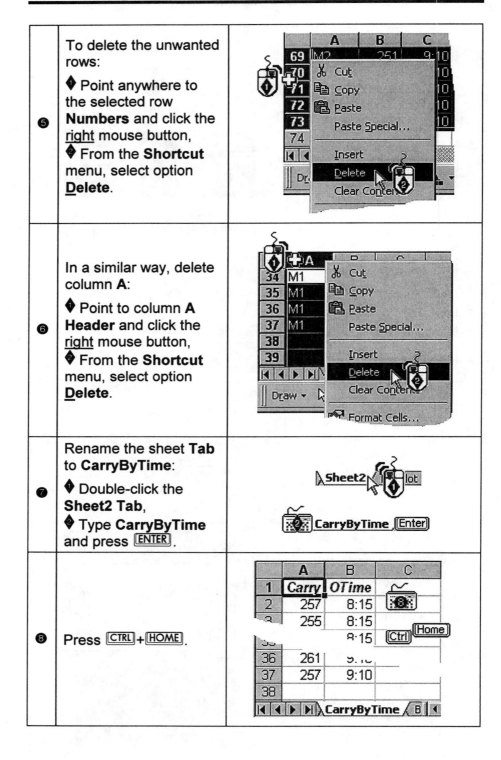
⑥	In a similar way, delete column **A**: ♦ Point to column **A Header** and click the <u>right</u> mouse button, ♦ From the **Shortcut** menu, select option **Delete**.	
⑦	Rename the sheet **Tab** to **CarryByTime**: ♦ Double-click the **Sheet2 Tab**, ♦ Type **CarryByTime** and press ENTER.	
⑧	Press CTRL + HOME.	

Now, you can compute the average and standard deviation for the three subsets of *Carry* (*OTime* 8:15, 8:45, and 9:10), using the **Data | Subtotal** command. Note that the data is already in the right format (sorted by *OTime*).

Procedure 6.6   Generating the Average and Standard Deviation from a Sorted List

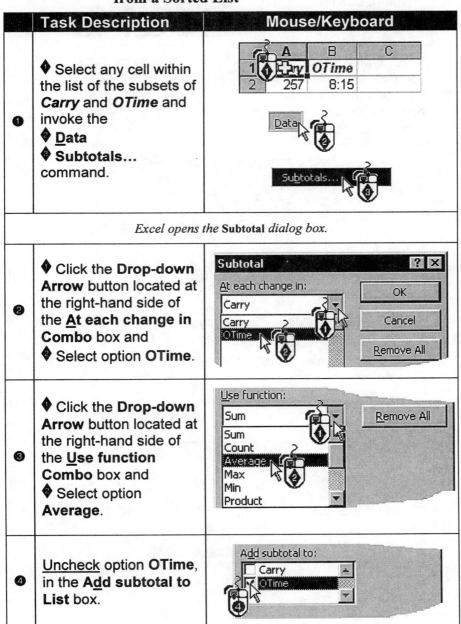

Note:

Sort Ascending

To sort a table by one column, select any cell in that column and click the **Sort Ascending** *or* **Sort Descending** *button.*

Sort Descending

Note:

*Before you do setp❼,
make sure that the
following options are
selected:*

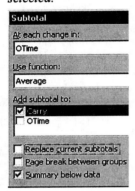

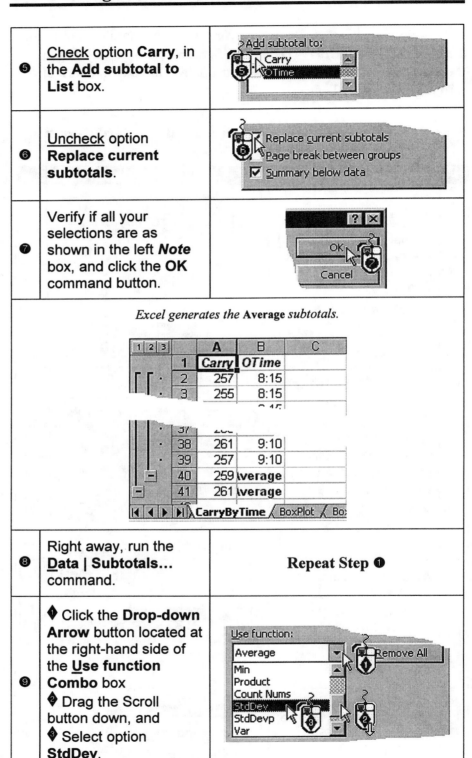

❺	<u>Check</u> option **Carry**, in the **A**dd **subtotal to** **List** box.	
❻	<u>Uncheck</u> option **Replace current subtotals**.	
❼	Verify if all your selections are as shown in the left *Note* box, and click the **OK** command button.	

Excel generates the **Average** *subtotals.*

❽	Right away, run the **D**ata	**S**ubtotals... command.	**Repeat Step ❶**
❾	◆ Click the **Drop-down Arrow** button located at the right-hand side of the **U**se function **Combo** box ❷ Drag the Scroll button down, and ◆ Select option **StdDev**.	

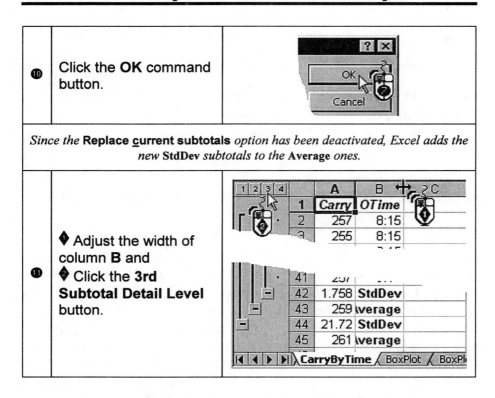

| | Click the **OK** command button. | |

Since the **Replace current subtotals** *option has been deactivated, Excel adds the new* StdDev *subtotals to the* Average *ones.*

| | ♦ Adjust the width of column **B** and
♦ Click the **3rd Subtotal Detail Level** button. | |

Figure 6.3 shows the average (**Average**) and standard deviation (**StDev**) subtotals for *OTime* = 8.15, 8:45, and 9:10, as well as their grand totals.

	A	B	C
1	*Carry*	*OTime*	
14	1.642	8:15 StdDev	
15	255.2	8:15 Average	
28	1.782	8:45 StdDev	
29	258.1	8:45 Average	
42	1.758	9:10 StdDev	
43	259	9:10 Average	
44	2.359	Grand StdDev	
45	257.4	Grand Average	
46			

Figure 6.3   Average and Standard Deviation Sub- and Grand Totals with Hidden Detail Data

Note:

You can use the **Plus** *and* **Minus** *buttons*

to reveal or hide the detail data.

You may wish to copy the measures to another location and then eliminate the outline view by executing the **Data** | **Subtotals** | **Remove All** command. Since the subtotals are calculated by the **Subtotal( )** function, all the measures would have to be copied (**Edit** | **Copy**) and then pasted using the **Edit** | **Paste Special** | **Values** command. Notice also that you would have to select each measure individually (via CTRL+🖰).

Exercise 11: Do you think the standard deviations of *Carry* for the *Model* M1 and the three times (*OTime* 8:15, 8:45, and 9:10) are significantly different?

Section 6.5 Descriptive Statistics Tool Versus Excel Functions.

In the preceding section you learned how to apply the **Descriptive Statistics** command provided by the **Add-In** based **Data Analysis Toolpak**. It is a convenient way to produce the basic summary measures for data organized in the form of a compact table. Each column of the table is expected to represent one variable (data set). The summary measures can also be produced by Excel standard functions. These do not impose any particular requirements or restrictions as to how the data sets are to be organized. For example, you can calculate the average for a data set scattered all over the workbook or even residing on many different workbooks. Needless to say, the functions are always there. They do not require any special installation procedures. More over, with those functions, you can develop your own reusable applications that will only require that you supply new data. The summary measures will be recalculated automatically.

In order to see if the **Data Analysis** tools and your formulas produce the same results, copy the summary measure from sheet **SummaryOfCarry** (range A1:B15) and paste them onto sheet **Sheet3** (starting at **A1**). Next, stretch column **A**, to reveal all descriptive labels, and then enter the formula as shown in Figure 6.4. Since some of the formulas located in higher rows (higher—visually, not numerically) depend on those in the lower rows, enter the formulas in the order from the bottom up.

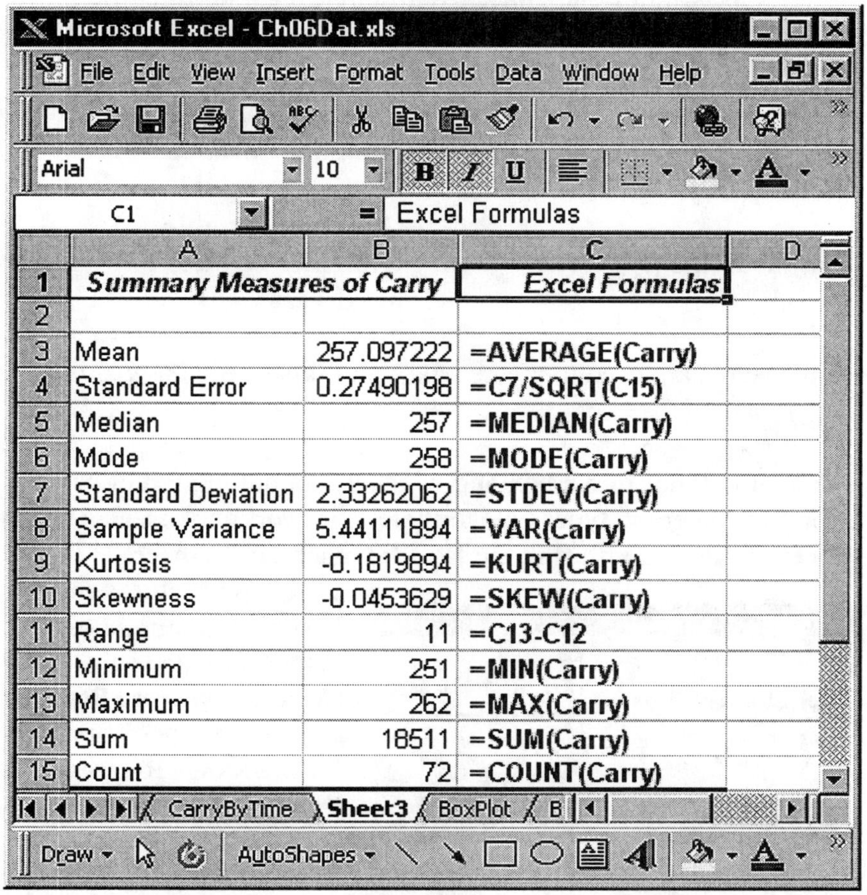

	A	B	C
1	*Summary Measures of Carry*		*Excel Formulas*
2			
3	Mean	257.097222	=AVERAGE(Carry)
4	Standard Error	0.27490198	=C7/SQRT(C15)
5	Median	257	=MEDIAN(Carry)
6	Mode	258	=MODE(Carry)
7	Standard Deviation	2.33262062	=STDEV(Carry)
8	Sample Variance	5.44111894	=VAR(Carry)
9	Kurtosis	-0.1819894	=KURT(Carry)
10	Skewness	-0.0453629	=SKEW(Carry)
11	Range	11	=C13-C12
12	Minimum	251	=MIN(Carry)
13	Maximum	262	=MAX(Carry)
14	Sum	18511	=SUM(Carry)
15	Count	72	=COUNT(Carry)

Figure 6.4 Summary Measures of Carry, Generated by the Data Analysis Tools (Column B) and Excel Formulas (Column C)

The **Data Analysis** tools do not provide all the essential summary measures. Two important measures are missing. They are discussed in the next section.

Section 6.6 Creating a Box Plot and Detecting Outliers

In the preceding sections you learned that you could decide if a particular observation was an outlier by looking at its **z**-score. A tool that gives you a visual way to see outliers is the box plot.

The descriptive statistics also provides what is sometimes referred to as the "5 number summary" of the data. These 5 numbers are:

▪	Minimum	=**Min**(*variable*)
▪	First Quartile (Q1),	=**Quartile**(*variable* , 1)
▪	Median (Q2),	=**Median** (*variable*)
		=**Quartile**(*variable* , 2)
▪	Third Quartile (Q3),	=**Quartile**(*variable* , 3)
▪	Maximum	=**Max**(*variable*)

Note:

Procedure 5.3 (page 105) shows step-by-step how to run a macro.

These measures are used to create a box plot (also called a *box and whisker plot*).

You will not find a box and whisker plot among Excel's rich repertoire of the graph types. Nevertheless, in workbooks **Ch06Dat.xls** and **MacDoIt.xls** you will find the **BoxPlot** macro. This macro command is stored in a Visual Basic module. When activated (via **Tools | Macro | Macros | BoxPlot | Run**), the macro first asks for the **Range Reference (Name)**. Since all our column variables are named, type **Carry** and **okay** this operation. The macro will draw a box and whisker plot based on the "5 number summary" of the data (Figure 6.5).

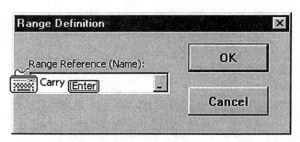

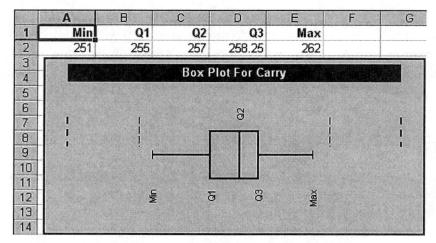

Figure 6.5 Box Plot with Outlier Fences for *Carry*

If you inspect the cells containing the **5** measures, you will see the following functions:

A1 =Min(Carry),
B1=Quartile(Carry,1),
C1=Quartile(Carry,2),   same as **=Median(Carry),**
D1=Quartile(Carry,3), and
E1=Max(Carry).

Notice that the first, second (median), and third quartiles are used to create the box. A vertical line is drawn across the box at the median (**Q2**). The left edge of the box is at the first quartile (**Q1**), and the right one is at the third quartile (**Q3**) value. The whiskers are the lines that extend from the left side and from the right side of the box. They go as far as the lowest (**Min**) and highest (**Max**) data values. The box and whiskers tell us more about the middle, dispersion, skewness, and flatness of the data set. The middle is provided by the median. The dispersion is explained by the interquartile range (**IQR = Q3-Q1**). For example, the smaller the **IQR** range compared to the total range (**Max - Min**), the smaller dispersion or variability and the smaller flatness. The position of the median (**Q2**) relative to the quartiles (**Q1** and **Q3**) determines skewness of the data set. The above plot shows the median a little closer to the third quartile (**Q3**) which indicates a slight negative skewness.

The **5 number measures** are frequently used to construct simple criteria for detecting potential outliers. Two types of limits are used to identify outliers: the so called *inner fences* and *outer fences*. The inner fences are **1.5** interquartile range distant from the quartiles:

> **Lower Inner Fence = Q1 - 1.5(IRQ)**
> **Upper Inner Fence = Q3 + 1.5(IRQ)**

The outer fences are **3** interquartile range distant from the quartiles:

> **Lower Outer Fence = Q1 - 3(IRQ)**
> **Upper Outer Fence = Q3 + 3(IRQ)**

Those numbers that fall between adjacent inner and outer fences are referred to as *mild* outliers. Those that are below the lower outer fence or above the upper outer fence are called *extreme* outliers. Both types of the outliers require further investigation to determine why they are so different from the rest of the data set. For a number to be eliminated from a variable (data set), it has be judged as an unusually different, or using the language of *probability*—unlikely small or large.

The box plot generated by the **BoxPlot** macro also shows the outlier fences.

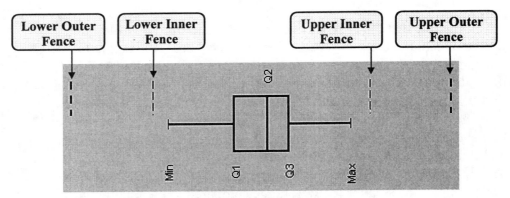

Figure 6.6 **Outlier Fences for *Carry***

One way to find out if there are potential outliers in a data set is to test the fence values against the set's minimum and maximum. The above box plot (for ***Carry***) shows that there are no outliers, since the lower fences are below sample minimum and the upper fences are above sample maximum.

Exercise 12: To see a different situation, run the **BoxPlot** macro for the *Wgt* variable.

Have you gotten the following box plot?

Note:

*The variable **Wgt** has many repetitive values, which is why **Q2** and **Q3** happen to be identical.*

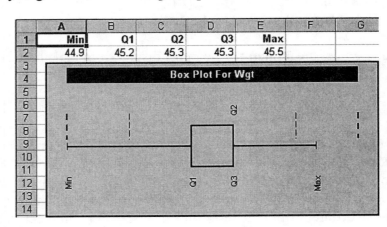

You know the variable *Wgt* may have some outliers. You can find out specific values of the outliers by running the **Outliers** macro. Like the **BoxPot** macro, this one can be executed from anywhere in the workbook. It inserts a new sheet, calculates the quartiles, minimum, maximum, and performs the test. The **Outliers** macro lists all detected outliers (if any) along with their positions in the data set.

Again, to run the macro, first invoke the **Tools | Macro** command. From the **Macro Name** list, select the option **Outliers** and choose the **Run** command. In the **Range Reference (Name)** input box, type **Carry** and press ⌈ENTER⌋ or click the **OK** command button.

The macro opens In the **Range Definition** dialog box (same as shown on page 158). In the **Range Reference (Name)** input box, type the name **Wgt** and click **OK** or press ⌈ENTER⌋. The final result is shown in Figure 6.7.

	A	B	C	D	E	F
1						
2		\multicolumn{5}{c}{**Outlier Report for Wgt**}				
3		Min	44.9		IRQ	0.1
4		First Quartile	45.2		Lower Outer Fence	44.9
5		Median	45.3		Lower Inner Fence	45.05
6		Third Quartile	45.3		Upper Inner Fence	45.45
7		Max	45.5		Upper Outer Fence	45.6
8		\multicolumn{2}{c}{**Mild Outliers**}			\multicolumn{2}{c}{**Extreme Outliers**}	
9		Position	Value		Position	Value
10		5	45.5		16	44.9
11		29	45.5		40	44.9
12		53	45.5		64	44.9

Figure 6.7   Outlier Report for the Variable Wgt

Exercise 13: Change the first value in the *Carry* set to **300** and see the macro **Outlier** outcome. Then change the value back to **257**.

Section 6.7 Investigative Exercises

In this section you will be asked to generate the appropriate descriptive statistics in order to answer the questions about the golf ball design data file. Remember that the numbers themselves must be interpreted within the context of the specific application.

1. Complete the following table:

Variable	Mean	Median	Range	Variance	Std.Dev
M1 ball All times					
Carry					
Tot Dist					
M2 ball All times					
Carry					
Tot Dist					

2. Change the values of *Carry* and *TotDist* for **Ball** 1 to **357** and **370** respectively. Redo the table with these new values.

Variable	Mean	Median	Range	Variance	Std.Dev
M1 ball All times					
Carry					
Tot Dist					
M2 ball All times					
Carry					
Tot Dist					

3. Compare the measures of the mean and the median found in exercise 1 with those calculated in exercise 2.

a) What has happened to the mean values as a result of having an outlier in the data set?

b) What has happened to the median values as a result of having an outlier in the data set?

c) What can you conclude about how the mean and the median respond to outliers?

d) How could you alter the calculation of the mean to eliminate this sensitivity to outliers?

---

4. Study the measures of variability found in exercises 1 and 2.

a) What has happened to the values of the range as a result of having an outlier in the data set?

b) What has happened to the values of the sample variance and standard deviation as a result of having an outlier in the data set?

c) What can you conclude about the usefulness of the range as a measure of variability?

---

5. Study the relationship between the mean and the median.

a) Find a variable such that the mean and the median are very close (or equal) to each other.   Examine the histogram of this variable. What type of shape does it have?

b) Find a variable such that the mean is larger than the median. Examine the histogram of this variable. What type of shape does it have?

c) Find a variable such that the mean is smaller than the median. Examine the histogram of this variable. What type of shape does it have?

d) Based on these observations what can you conclude about the shape of the distribution based on the relationship between the mean and the median?

---

6. Compare the mean of the variable *Carry* for ball *Model* **M1** @8:15AM to the mean of the variable *Carry* for ball *Model* **M1** @8:45AM. Notice that their means are very close to each other. Now compare the 2 standard deviations.

a) What do you notice?

b) What does this tell you about the adequacy of only using the mean to describe a data set?

c) Look at a histogram of each of these variables. What can you learn from the graphs about how the two variables differ?

---

7. Further investigate the variables used in exercise 5.

a) Complete the following table:

Variable	OTime	Ball (#)	Distance		
			Total Yards	From The Mean Measured In	
				Yards	Standard Deviations
Carry	8:15	13	256		
Carry	8:45	41	256		
Carry	8:15	14	255		
Carry	8:45	42	255		
Carry	8:15	16	257		
Carry	8:45	40	257		

b) What is another name for the values you calculated for the last column?

c) What do you notice about the values in column 5? What does this tell you about the usefulness of simply looking at the distance between an observation and its mean?

---

8. Is it possible for two variables to have similar variances but vastly different means?

a) Find two such variables in the data set and create a histogram for each variable.

b) Compare the histograms. If you know that two variables have similar variances but different means, what does that tell you about the behavior of the two variables?

---

9. Consider the variable *TotDist* for the **M2** ball for all times.   Look at how well the data for this variable fits the empirical rule. Recall that the empirical rule states that **68%** of the data will fall within **1** standard deviation of the mean, **95%** of the data will fall within **2** standard deviations of the mean, and **99%** of the data will fall within **3** standard deviations of the mean.

a) Display the data as a histogram.

b) Complete the following information (from exercise #1):

Count	
Mean	
Standard Deviation	
Mean - 1 StDev	
Mean + 1 StDev	
Mean - 2 StDev	
Mean + 2 StDev	
Mean - 3 StDev	
Mean + 3 StDev	

c) Mark the last 6 values on the histogram.

d) How many data values fall within **1** σ of the mean? _____
Divide this by the number of observations for the variable (N) to get a
%: _____

How many data values fall within **2** σ of the mean? _____
Divide this by the number of observations for the variable (N) to get a
%: _____

How many data values fall within **3** σ of the mean? _____
Divide this by the number of observations for the variable (N) to get a
%: _____

e) How closely do the percentages found in part d) match the
empirical rule? What does this tell you about the general shape of the
graph?

f) Consider the value for *Ball 2* . Mark it's location on the x-axis of
the histogram. Where does it fall relative to the markings you made in
part c)?

   Now find its z-score. What then does the z-score tell you about the
data point?

10. Consider variables with hardly any variability.

a) Can you find a variable in the data set that has a zero (or close to zero) variance?

b) What can you say about the observed values for this variable?

---

11. Compare the mean for the variable *Carry*, *Model* **M1** at **8:45AM** to the mean of the variable *Carry*, *Model* **M2** at **8:45AM**. Notice that they are close in value but not exactly the same.

a) If you observed **12** more values of *Carry* for the **M1** design at **8:45AM** and calculated the mean of those **12** values, would you get exactly **258.0833**?   Would the second mean be close to **258.0833**? Why or why not?

b) If you observed **12** more values of *Carry* for the **M2** *Model* at **8:45AM** and calculated the mean of those **12** values, would you get exactly **257.5833**?   Would you expect it to be close to **257.5833**? Why or why not?

c) What does this tell you about the sample means from sample to sample?

d) Now, considering your answer to part c), is it possible that the true underlying population mean of the variable *Carry* for both designs at **8:45AM** is the same despite the fact that you have observed **2** slightly different sample means? Why or why not?

Chapter 7   "Credit Problems?"

A Study of The Binomial Distribution

Section 7.1 Overview

In the last three chapters you have learned how to *describe* your data set. When you construct a histogram of your data, what you are looking at is the <u>distribution of the sample </u>data. However, we must remember that it is NOT the sample which we are ultimately interested in but rather the population from which the sample was drawn. Thus, although it is quite helpful to view the distribution of the sample, what we really would like to know is the <u>distribution of the population</u> from which this sample was drawn. You can think of the sample as evidence upon which you wish to draw some *inferences* about the population. In a sense you must be a bit of a detective.

In order to successfully detect the underlying population distribution you must know a little bit about the behavior of some of the distributions which are commonly found. There are two basic classes of distributions: discrete distributions and continuous distributions. This chapter will focus on one of the most commonly used discrete distributions, the **BINOMIAL DISTRIBUTION.** In Chapter 8, you will investigate the most commonly used continuous distribution, the Normal Distribution.

Statistical Objectives: After reading this chapter and doing the exercises a student will:

- Know how to detect a Binomial Variable.
- Know how changing the values of n and p affects the shape of the Binomial Distribution.
- Be able to makes some statements about the likely value of the parameter p, the probability of success.
- Know how to compare two binomial variables.

Section 7.2 Problem Statement

In recent years the economy has experienced a recession. Banks and other financial institutions have imposed more stringent conditions on companies trying to get loans. This is clearly to protect the banks from making "bad" loans. However, as a result of this conservative policy, many small companies have been forced to close or reduce their size because they can not meet the stiffer requirements needed to get a loan. A medium size New England city has been very concerned about this problem and has tried to find out the specific nature of the credit problems facing small businesses. They have done this by administering a questionnaire every 6 months.   The questionnaire consists of many questions, which fit the model of the binomial distribution.

The survey was mailed to 1536 companies within a 10 mile radius of this city. A total of 166 usable responses were received. On the basis of your analysis of these 166 responses you will be asked to describe the nature of the credit problem to the Chamber of Commerce.

Section 7.3 Characteristics of the Data Set

FILENAME:	Ch07Dat.xls	An Excel Workbook
SIZE:	COLUMNS	11
	ROWS	166

Figure 7.1 shows the first 10 rows of the actual data file.

Notes on the data set:

For all variables, the code of zero (0) was used when the question was not answered by the respondent.
1. The variable *Number* simply numbers the respondents from 1 to 166.

	A	B	C	D	E	F	G	H	I	J	K
1	Number	Size	Employees	Nature	Problem	Understands	Concerned	Call	Loan	Collateral	Access
2	1	2	2	1	1	2	1	2	2	0	2
3	2	1	2	3	1	2	2	0	2	2	1
4	3	4	3	1	2	1	2	2	2	2	0
5	4	1	1	1	2	1	2	2	2	2	2
6	5	1	2	1	2	0	0	0	0	0	0
7	6	3	1	5	2	0	2	2	2	2	1
8	7	3	3	2	2	1	1	2	2	2	1
9	8	2	2	3	2	1	2	2	2	2	2
10	9	3	5	2	2	1	2	2	2	1	2

Data / Sheet2 / Sheet3 / Sheet4 / Sheet5

Figure 7.1   First 10 Rows of Ch07Dat.xls Data

2. The variable *Size* refers to the annual sales of the company and is coded as follows: 1   Under $1 million
    2      $1- 5 million
    3      $5- 10 million
    4      $11- 20 million
    5   Over $20 million

3. The variable *Employee* refers to the number of employees which the company currently employs. This variable was coded as follows: 1   0-5 employees
    2   6-10 employees
    3   11-50 employees
    4   51-150 employees
    5   151-250 employees
    6   Over 250 employees

4. The variable *Nature* refers to the nature of the business and is coded as follows:
    1   Manufacturing
    2   Retail
    3   Service
    4   Real Estate
    5   Other

The coding scheme for the remaining variables is as follows:
    1   **Yes**
    2   **No**

5.   The variable *Problem* contains the respondents answer to the question: "Are you experiencing recession related problems?"

6.   The variable *Understd* contains the respondents answer to the comment: "Bank understands my business".

7.   The variable *Concern* contains the respondents answer to the comment: "Concerned my note may be recalled".

8.   The variable *Call* contains the respondents answer to the comment: "Bank is planning to call my loan".

9.   The variable *Loan* contains the respondents answer to the comment: "Bank has called my loan".

10.   The variable *Collater* contains the respondents answer to the comment: "Bank has demanded more collateral".

11.   The variable *Access* contains the respondents answer to the comment: "Access to credit is effecting my business".

Open

Click the **Open** *button or use the* **File** | **Open** *command (*ALT*+*F*,* O*) to open the file.*

Open the data set named **Ch07Dat.xls** from your data disk. Note that Procedure 3.1, on page 33, contains detail instruction about how to open an Excel workbook.

Section 7.4 Detecting a Binomial Variable

In order to analyze this data set, you must first look at whether or not any of these variables fit the model for a Binomial Random Variable. Recall that a Binomial Random Variable has the following properties:

1   The experiment consists of **n** independent trials;
2   Each trial results in one of two possible outcomes; a successful outcome or a failure outcome;
3   The probability of a successful outcome is the same from trial to trial and is called **p**;
4   The probability of a failure outcome is the same from trial to trial and is called **q = 1-p**.

Consider the variable *Problem* as described above. Do you think this variable fits the model for a Binomial Random Variable? Let's have a look!

The "experiment" consists of **166** independent responses to the question: "Are you experiencing recession related problems?". They can be considered independent unless we have reason to believe that the respondents influenced each other in answering the questionnaire. There is no reason to believe this. So each response to this question is one "trial" and we have met the first criteria.

The possible responses to this question were either **Yes** or **No**. If we consider the response "**Yes**" as the successful outcome and the response "**No**" as the failure outcome, then we have met the second criteria.

The probability of a successful outcome is then the probability that the response is **Yes** from any random respondent. This probability is unknown. In order to satisfy the third and fourth criteria, you must see if the value of **p** (and therefore **q**) is the same from trial to trial.   If we did know **p**, it would be equal to the percentage of the **1536** members of the population who are experiencing recession related problems. Pretend for a moment that you do know the value of **p**. Then if you randomly selected a respondent from the population, the chance that it would be experiencing recession related problems would be **p**. If you made another random selection, would the chance that it was experiencing recession related problems still be **p**? What do you think?

Unless you sampled with replacement, that is you allowed the first respondent to be picked again, the answer would have to be no, the value of **p** would not be the same as on the first pick. Thus we are in violation of criteria **3** and **4**. In most actual situations you will not sample with replacement as it does not make much sense to do this. So technically speaking you almost never have a true binomial random variable. In this case we can just skip the rest of this chapter, right? No! This is the first instance in this workbook where we must use our understanding of statistics to decide if we can "bend the rules a bit and still be OK". This will not be the last time we encounter this. In actual data situations you almost never meet all of the criteria or assumptions of the theory. Thus, it is vitally important that you understand the theory well enough to decide what rules you can and cannot bend!

In this case we must decide if sampling without replacement yields a situation where the value of **p**, although changing from trial to trial, is not changing by much. Suppose for example that there are **231** out of **1536** companies experiencing recession-related problems. Then on our first selection, the chance of picking someone who is experiencing a recession related problem is **231/1536 = .15039**. On our second selection, the probability of selecting someone who is experiencing a recession related problem is either **231/1535= .15048** or **230/1535=.14983**. As you can see, the value of **p** is changing but not by much.   So we accept the slight deviation from the exact model and use the binomial distribution to model the variable ***Problem***.

Exercise 1. What other variables in this data set can be considered binomial variables? For each variable, identify the successful outcome, the failure outcome and the value for **n**.

Now that we have decided to use the binomial distribution, we must decide how to use it, since we do not know the value of **p**. In fact, your job is to make some statements about the likely value of **p** for each of the binomial variables. In order to do this we need to have a good understanding of what the binomial distribution looks like and how it behaves. We will use Excel to save us the tedious calculations.

Section 7.5 The Binomial Distribution in Excel

The first step in generating a binomial distribution in Excel is to create a column of numbers, which gives all of the possible values of the binomial random variable. Remember that the binomial random variable, **X**, is defined to be the total number of successes in **n** trials. Thus **X** can be **0, 1, 2, 3**, and so on all the way up to **n**, where **n** refers to the total number of trials in the experiment. This is the sample size. Thus, in order to tabulate the entire distribution you would start at **0** and end at **n**. After all you can't have less than zero successes or more than **n** successes in a sample of size **n**. If we continue to examine the variable ***Problem*** in this data set, then **n = 166**. Having the random variable X values, we will compute their Binomial probabilities and cumulative probabilities.

Procedure 7.1 Generating Binomial Probabilities

	Task Description	Mouse/Keyboard
❶	Switch to **Sheet2** and rename its **Tab** to **BinProb**.	
❷	Type and format the labels and numbers as shown on the right. If you hate typing, run the **GetBinProbHeader** macro.	*Note:* Procedure 5.3 shows how to run a macro command.
❸	Fill range **A5:A171** with the series of numbers **0,1,…,166**. *Hint*: Type **0** in **A5** and press CTRL + ENTER . Then execute this command: **Edit	Fill	Series |** **⊙ Columns |** **Step value: 1 |** **Stop value: 166 |** **OK**.	
❹	Name all the **x** values (**A5:A171**) as **x**.	

Note:
*If you choose the macro way, you will only need to enter numbers **0.3** and **166** in cells **B2** and **B3**, respectively.*

Note:
Procedure 5.12, on page 124, shows how to fill a range with an arithmetic series.

Note:
Procedure 5.5, on page 107, shows detail steps about naming a range.

Note:
Procedure 5.1, on page 100, shows detail steps about naming ranges via the **Insert | Name | Create** *command.*

⑤ Also name
B2 as **p**
and
B3 as **n**.

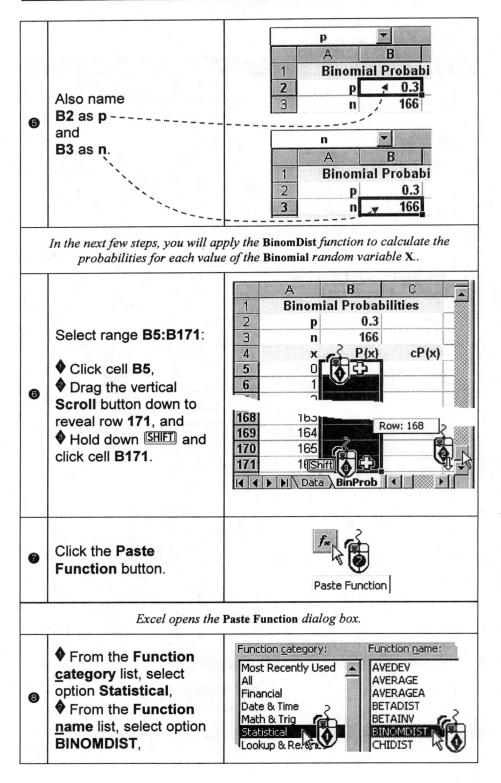

In the next few steps, you will apply the **BinomDist** *function to calculate the probabilities for each value of the* **Binomial** *random variable* **X**.*

⑥ Select range **B5:B171**:

❶ Click cell **B5**,
❷ Drag the vertical **Scroll** button down to reveal row **171**, and
❸ Hold down ⌈SHIFT⌋ and click cell **B171**.

⑦ Click the **Paste Function** button.

Paste Function

Excel opens the **Paste Function** *dialog box.*

⑧ ❶ From the **Function category** list, select option **Statistical**,
❷ From the **Function name** list, select option **BINOMDIST**,

	Excel opens the **BINOMDIST FUNCTION WIZARD** *dialog box.*
⑨	◆ In the **Number_s** input box, type **x** and press TAB, ❷ In the **Trials** input box, type **n** and press TAB, ❸ In the **Probability_s** input box, type **p** and press TAB, ◆ In the **Cumulative** input box, type **False**, hold down CTRL, and press ENTER.
⑩	In order to generate the cumulative distribution in **C5:C171**, you can adopt the steps ❽-⑬ of **Procedure 5.7 (page 112)** or apply the steps ❼-❾ of this procedure to the **BINOMDIST(x,n,p,True)** function.

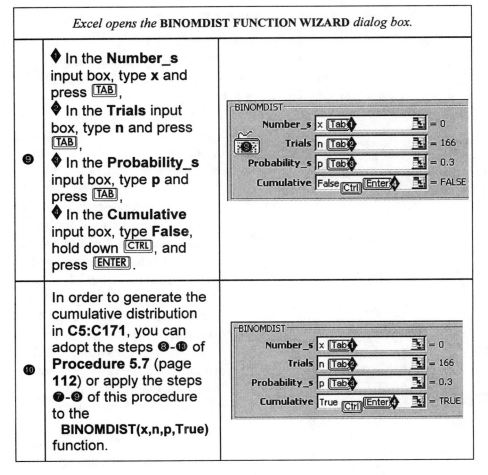

Note:
As you define the references, the **Function Wizard** *shows what they are equal to.*

Note:
Recall, when you press CTRL + ENTER, *in order to enter a formula, Excel replicates the formula within the selected range. So it is a shortcut for* **Edit | Copy** *and* **Edit | Paste**.

Note:
If you decide to use, the **BINOMDIST(x,n,p,True)** *function, make sure that you type* **True** *in the* **Cumulative** *input box.*

The **BinomDist** function takes four arguments: the number of successes (**x**), the total number of trials (**n**), the probability of success on each trial (**p**), and a flag or switch argument (**s**) ,

BinomDist(x,n,p,s).

In order to use this function to calculate the probability of **x** successes in a series of **n** trials, the switch, **s**, is set to **False**. For the probability of no more than **x** successes in a series of **n** trials, the switch, **s**, is set to **True**. If we let X_p^n represent the number of successes in a series of *n* trials, then we have

$$P(X_p^n = x) = \text{BinomDist}(x,n,p,\text{False}),$$

$$P(X_p^n \le x) = \text{BinomDist}(x,n,p,\text{True}).$$

For example, **BinomDist(5,10,0.3,False)** returns the probability of having $X^{10}_{0.3} = 5$ successes in **n = 10** trials, if the probability of success in each trial is **p = 0.3**, whereas **BinomDist(2,10,0.3,True)** determines the probability of having up to **2** successes (**0,1**, or **2**) also in a series of **n = 10** trials where **p = 0.3**.

	A	B	C
1	Binomial Probabilities		
2	p	0.3	
3	n	166	
4	x	P(x)	cP(x)
5	0	1.93E-26	1.9332E-26
6	1	1.38E-24	1.3947E-24
7	2	4.86E-23	5.0022E-23
50	45	0.049631	0.234...
51	46	0.05595	0.29084964
52	47	0.061222	0.35207146
53	48	0.065048	0.41711964
54	49	0.067134	0.48425391
55	50	0.067326	0.55157999
56	51	0.065629	0.61720878
57	52	0.062203	0.67941189
58	53	0.057341	0.73675276
59	54	0.051425	0.78817751
	55	0.04488	0.83305729
169			
170	165	6.17E-85	1
171	166	1.59E-87	1

Figure 7.2   Fragment of Binomial Probabilities (p=0.3, n=166)

Figure 7.2 shows a fragment of the results.

Column **x**, gives the values of the random variable. Since we are considering the variable *Problem*, this column corresponds to various possible values for the number of respondents who answered **Yes** (the successful outcome) to the question: "Are you experiencing recession related problems?". Clearly anywhere from **0** to **166** might have responded **Yes** to this question.

Column **P(x)**, gives the probability corresponding to the value shown in the **x** column. For example, if you look at the row where **x** equals **50**, then you will see the probability is **0.067326**. This is the probability you would get if you plugged the values **x = 50, n = 166** and **p = 0.3** directly into the formula for the binomial distribution, **BinomDist(50,166,0.3,False)**. As you examine column **P(x)**, you will see lots of zeros and small probabilities at the top and bottom of the column. This is because it is unlikely that you would observe a relatively few number of **Yes** answers out of **166** and it is also unlikely that you would observe mostly **Yes** responses when **p = 0.3**. In fact, you have probably noticed that most of the non-zero probability lies in the range **20<X<60**. You can tell this by observing the tiny probabilities in the second column for values of **X** outside this range.

Exercise 2. What is the probability of observing exactly **45 Yes** responses for the variable *Problem*, if **p = 0.3**?

You should notice that the values in the cumulative probability column are increasing. The third column, labeled **cP(x)**, gives the cumulative probability distribution function. For the value of **X = 50** the cumulative probability function is **P(X ≤ 55)** and is given in the table as **.55157999**. This means there is about an **55.2%** chance that you would see at most **50 Yes** responses for the value of **p** equal to **0.3**.

Exercise 3. Using the information shown in Figure 7.2, find the probability of observing **53 Yes** responses, if **p** is actually **0.3**.

Exercise 4. Find the probability of observing **50** or less **Yes** responses, if **p** is **0.3**.

Exercise 5. Find the probability of observing more than **50 Yes** responses, if **p** is **0.3**.

It is a good time to save your workbook.

Section 7.6 Plotting the Binomial Distribution in Excel

What does the graph of this binomial distribution look like? To display a graph of the distribution shown in the column labeled **P(x)**, you will develop an **XY** chart. You learned the basics of creating charts in Chapters 4 and 5. Here are just a few hints that should help you create the chart without having you to go back to those chapters.

Procedure 7.2   Plotting the Binomial Probability Distribution, Using an XY Chart

	Task Description	Mouse/Keyboard
❶	On the **BinProb** sheet, select range **A5:B171**.	[🖱️] [F5] A5:B171 [Enter]
❷	Invoke the **Chart Wizard**.	Chart Wizard

Note:
You can use the

[F5]

function key (a shortcut for **Edit** | **Go To...**) *to quickly jump to a cell or to select a range via keyboard.*

Note:
Procedures 4.2-4.6 on
pages 55-60 show
how to format a chart.

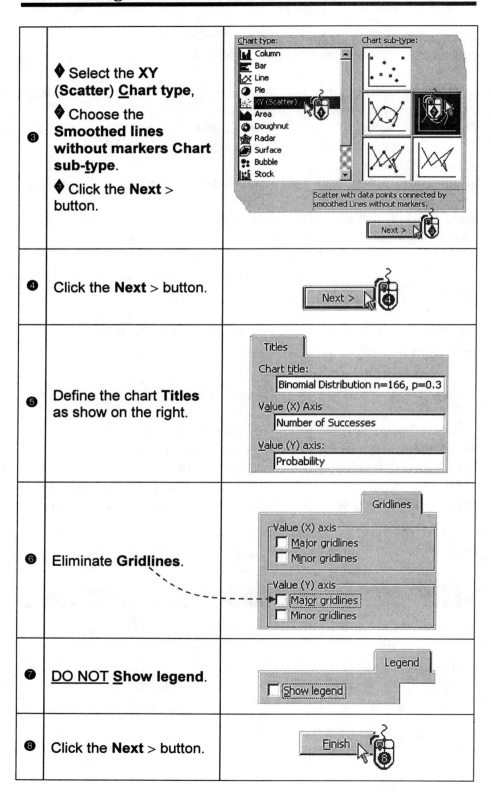

❸	◆ Select the **XY (Scatter) <u>C</u>hart type**, ◆ Choose the **Smoothed lines without markers Chart sub-<u>t</u>ype**. ◆ Click the **Next >** button.
❹	Click the **Next >** button.
❺	Define the chart **Titles** as show on the right.
❻	Eliminate **Grid<u>l</u>ines**.
❼	<u>DO NOT</u> <u>S</u>how legend.
❽	Click the **Next >** button.

When you have completed all the **Chart Wizard** dialog boxes and changed format of chart **Plot Area, Title, Value (X) axis, Value (Y) axis**, and **Series 1**, you will see a column chart like the one shown in Figure 7.3.

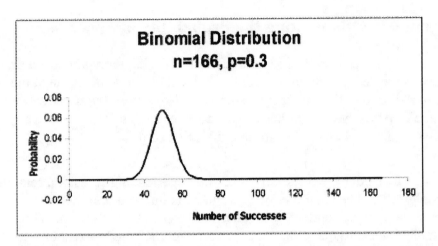

Figure 7.3   Binomial Distribution Function for n=166 and p=0.3.

Note:
You can easly change the **XY** *chart into a* **Column** *chart.*

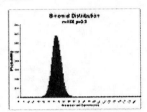

To do this, right-click **Chart Area.** *From the menu select* **option Chart Type.** *In the dialog box, choose* **the Column Chart type** *and* **OKay** *this operation.*

It is important for you to have an understanding of what the binomial distribution looks like However, it is equally as important to realize that if you construct a column chart of the responses to any of the binomial variables in this data set, it will NOT look like the graph shown in Figure 7.3, nor should it. They are completely different graphs. Let's see why this is so.

Exercise 6. Construct an column chart to display the responses stored in the variable **Problem**.

Notice that the chart simply gives a visual display of how many respondents answered **Yes, No** or gave **No Response** to the question: "Are you experiencing recession related problems?". This graph does not have as its **X** axis the number of successes in **166** responses as the graph in Figure 7.3 does. They are NOT comparable graphs. The graph of the binomial distribution shows how likely it is that you would get **x** successes if **p = 0.3**. The graph you made in **Exercise 6** shows that there were **25** successes (**Yes** responses) for this survey. Thus **X = 25** just for this survey. It corresponds to one of the **166** different possible values of **X**.

This is not to say that you should not create a chart of the responses to the survey. Rather it is important to understand that when you do so you should NOT expect it to have the shape of a binomial distribution even though it is a binomial variable.

Section 7.7 Estimating the Probability of Success, p

Much of our discussion thus far in the chapter has focused on how to create a binomial distribution table and graph when the value of **p**, the probability of success is known. However, if you knew the value of **p** you would not have to do a survey! This means that we must somehow be able to estimate the value of **p**.

Consider the question "Are you experiencing recession related problems?". The Chamber of Commerce is interested in the percentage of respondents who answered **Yes** to this question. Thus, it makes sense to make the "successful outcome" the **Yes** response. We could estimate **p** by taking the number of **Yes** responses and dividing by the total number of respondents.

Why do you think this might be misleading? Suppose of the **166** responses, there were **25 Yes's**, **45 No's** and **96** respondents who did not answer the question. Using the method just suggested would yield an estimated **p** value of **25/166 = .15**. This is misleading because of the **70** responses to this question **25** of them were **Yes** responses. Thus, a more accurate reflection of the survey result would be to estimate **p** as **25/70 = 0.36**. This is considerably different and gives a much different picture of the financial situation facing the companies in this New England city. Clearly you should do some investigation as to why there was such a high non-response rate to this question.

Note:
To name the data set, switch to **Data** *sheet, highlight the entire set (along with the top row holding the variable names) and run the* **Insert | Name | Create** *command.*
Check Procedure 5.1, page 100, for more help.

In order to estimate the value of **p** using Excel you need to have Excel count for you the number of **Yes** responses (coded **1**), the number of **No** responses (coded **2**), the number of respondents who did not answer the question (coded **0**). The following instruction provides all necessary steps. It assumes that our data set is named using the variable names (*Number*, *Size*, *Employee*,...).

Procedure 7.3 Using the CountIf( ) Function to Determine Proportions

	Task Description	Mouse/Keyboard
❶	Switch to **Sheet3**.	
❷	Rename the sheet as **ProblemProportions**.	
❸	Run the **GetProportionHeadings** macro (**Tools | Macro | Macros | MacroName: GetProportionHeadings | Run**).	
❹	In cell **C4**, enter function **CountIf(Problem,B4)**.	
❺	Copy the function.	
❻	Select range **C5:C6**.	
❼	Paste the function.	
❽	Select cell **C7**.	

Note:
If the data set is not named, the CountIf *function must use the absolute range reference for the variable* **Problem***:*
=COUNTIF(
$E$2:$E$167,B4)

Note:
Countif(Problem,B4)
takes the value from cell **B4** *and finds out how many times it occurs in the variable (range)* **Problem***.*

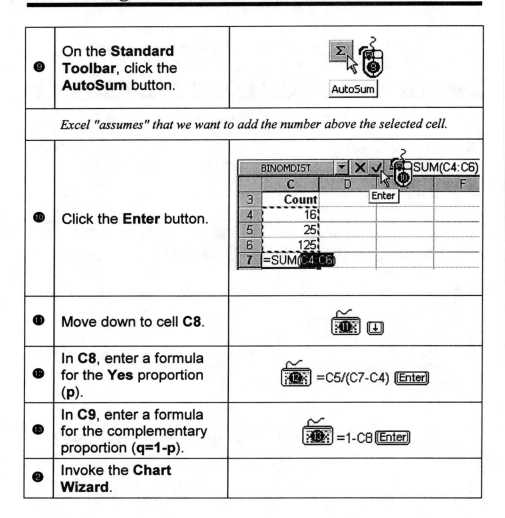

⑨	On the **Standard Toolbar**, click the **AutoSum** button.	
	Excel "assumes" that we want to add the number above the selected cell.	
⑩	Click the **Enter** button.	
⑪	Move down to cell **C8**.	
⑫	In **C8**, enter a formula for the **Yes** proportion (**p**).	=C5/(C7-C4) [Enter]
⑬	In **C9**, enter a formula for the complementary proportion (**q=1-p**).	=1-C8 [Enter]
⓿	Invoke the **Chart Wizard**.	

Figure 7.4 shows the results.

	A	B	C	D
1	**Summary Count for the *Problem* Variable**			
2				
3		**Variable Value**	**Count**	
4		0	16	
5		1	25	
6		2	125	
7		**Total**	166	
8		**Yes-Proportion, p**	0.166667	
9		**Not-Yes-Proportion, q**	0.833333	

Figure 7.4   Counts and Proportions for the
***Problem* Variable**

When you tell the Chamber of Commerce that **16.7%** of the respondents are experiencing recession related problems they might want to know if small businesses are being "hit" harder than larger businesses. In order to answer this question you need to construct what is known as a cross tabulations table. This is simply a table with the various possibilities for the variable *Problem* shown as the rows and the various possibilities for the variable *Size* shown as the columns.

Procedure 7.4   Generating a Cross Tabulation Report for Variables *Problem* and *Size*

Task Description	Mouse/Keyboard
❶ Switch to the **Data** sheet and select any cell within the data range.	
❷ Invoke the **Data | Pivot Table** Report command.	
Excel starts the **Pivot Table Wizard**.	
❸ Accept the default selection for the data source (⊙ **Microsoft Excel data list or database**).	
❹ Also accept the selected data **Range**.	Range: $A$1:$K$167

Note:
Instead of pressing ⟦ALT⟧+⟦R⟧, *you could drag the field and drop it onto* **R**OW.

Note:
Instead of pressing ⟦ALT⟧+⟦D⟧, *you could drag the field and drop it onto* **D**ATA.

Note:
Instead of pressing ⟦ALT⟧+⟦C⟧, *you could drag the field and drop it onto* **C**OLUMN.

❺	◆ Select (click) field **Problem** and ❷ Insert the field into the **ROW** section of the diagram (**Pivot Table**). ❸ Insert the field into the **DATA** section of the diagram (**Pivot Table**).	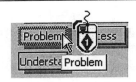 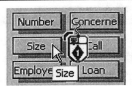
❻	◆ Select (click) field **Size** and ◆ Insert the field into the **COLUMN** section of the diagram.	

*Excel attempts to generate cross-tabulated sums. Since you want to count zeros, ones and twos for each **Size** category, you need to redefine the **D**ATA operation.*

❼	Double click the **Sum of Problem** button.	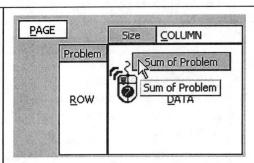

*Excel opens the **PivotTable Filed** dialog box.*

❽	Change the **Summarize by** function to **Count** by double clicking it.	

⑨	Back in the **PivotTable Wizard Step 3 of 4** dialog box, click the **Next >** button.	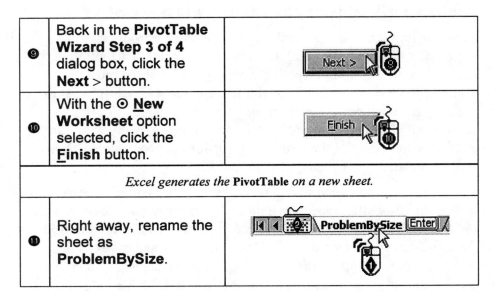
⑩	With the ⊙ **New Worksheet** option selected, click the **Finish** button.	
	*Excel generates the **PivotTable** on a new sheet.*	
⑪	Right away, rename the sheet as **ProblemBySize**.	

It is a good time to save your workbook.

Figure 7.5 shows the cross-tab (**PivotTable**) report.

	A	B	C	D	E	F	G	H
1	Count of Problem	Size						
2	Problem	0	1	2	3	4	5	Grand Total
3	0		7		4	2	3	16
4	1		6	7	5	3	4	25
5	2	1	47	14	33	16	14	125
6	Grand Total	1	60	21	42	21	21	166
7								

|◄ ◄ ► ►|\ **ProblemBySize** / Data / BinProb / ProblemProportions |◄|

Figure 7.5   Cross Tabulation of *Problem* by *Size*

From this **PivotTable** report, we can see that there are **53** (6+47) respondents Under $1 million in *Size* (*Size* =1) who answered either **Yes** (coded **1**) or **No** (coded **2**) to the *Problem* question. Of these **53** companies, **6** or **11%** are experiencing recession related problems.

Exercise 7. For each of the other *Size* categories, find the percentage of respondents who answered **Yes** to the *Problem* question. What can you conclude?

Section 7.8 The Beginnings of Hypothesis Testing

Although you have probably not yet been introduced to hypothesis testing, it is useful to study the binomial distribution with this technique in mind. Hypothesis testing is one of the most commonly used tools of statistical inference and it is helpful to start viewing data this way as early in the game as possible. For this reason, a preliminary glimpse of hypothesis testing is given in this section without any of the terminology or formal procedures. These will be discussed in Chapters 10, 11 and 12.

Note:
Symbolically, we use
p *to denote the true*
(population)
proportion. We use
$\bar{p}$ *(p hat) to*
represent an
estimate of **p**.

In addition to estimating the value of **p**, it is often necessary to answer the following type of question: *Is it reasonable to believe that the true value of p is 0.50?* If you had an estimate of *p* equal to **0.53** from the survey data, then it would be rather easy to say **yes** to this question. Now, suppose that $\bar{p} = 0.20$. Then the answer to this question is easily no. However, suppose $\bar{p} = 0.57$. Now, what is your answer to this question? It gets a little bit harder to decide if **0.50** is the true proportion of successes in the population given that you have observed **57%** successes in the sample. Remember that the sample is just a piece of the population. It is possible that the proportion of successes in the population is only **50%** but in your sample it was **57%**. Having established that it is possible, the right question is "how *likely* is it that we would observe a $\bar{p}$ of **0.57** IF the true proportion is **0.50**.

In order to answer this question, you need to use Excel to calculate the likelihood or probability of the observed sample, if **p** is really **0.50**. Again consider the variable *Problem* . Using the information that you found in the previous section, you know that there were **150** non-zero responses to this question and that **25** of these were **Yes** responses. You need to calculate the probability of observing **25** or fewer successes (**x**), in **150** trials (**n**), if the true value of **p** is **0.50**.

To find this probability, you could "reuse" the list of binomial probabilities created on sheet **BinProb**. Unfortunately, you would have to change the probability of success (**p**) there, which would also change the data list and the chart. Instead of doing this, you will generate a separate model that will accept the number of trials (**n**), the probability of success (**p**), and the number of successes (**x**), and then it

will produce the binomial probabilities, **P(X=x)** and **P(X≤x)**.
Procedure 7.5 shows how to generate the model.

Procedure 7.5   Generating the Binomial Probability Model

	Task Description	Mouse/Keyboard
❶	Switch to **Sheet4** and rename it as **BinomProbModel**.	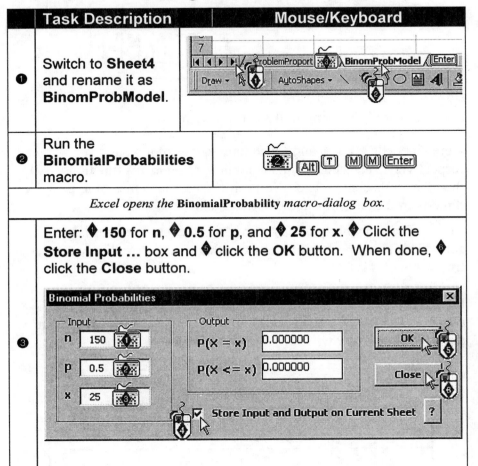
❷	Run the **BinomialProbabilities** macro.	

Excel opens the **BinomialProbability** *macro-dialog box.*

❸	Enter: ♦ **150** for **n**, ♦ **0.5** for **p**, and ♦ **25** for **x**. ♦ Click the **Store Input ...** box and ♦ click the **OK** button. When done, ♦ click the **Close** button.

Note:
If you check the **Store Input and Output on Current Sheet** *box on, the macro will show the output in the dialog box and it will also generate* **the Binomial Probability** *model on the current sheet. To see a warning message, click the* 🔲 *button.*

Given you did check the **Store Input and Output on Current Sheet** box **on,** you will see the following model on the current sheet.

=BINOMDIST(C5,C3,C4,FALSE)

E4		=	=BINOMDIST(C5,C3,C4,TRUE)	
A	B	C	D	E	F
1					
2		Binomial Probabilites			
3		n	150	P(X=x)	1.37E-17
4		p	0.5	P(X<=x)	1.71E-17
5		x	25		
6					

Figure 7.6   Binomial Probability Model

Note that cells **C3**, **C4**, and **C5** contain values of **n**, **p**, and **x**, respectively. The cells **E3**, **E5** contain the **BinomDist** functions. All the other cells contain text. You can change the values of **n**, **p**, and **x** directly in the cells or run the macro again.

The probability of observing **25** or fewer successes is close to zero. This small probability tells you that it is quite unlikely that the true value of **p** is **0.50**.

Exercise 8. For the variable *Problem*, find the probability of the observed sample **IF p = 0.10, 0.20, 0.30**. Summarize your results in the table shown below.

True **p** = Probability of Success	Probability of Observed Sample
.10	
.20	
.30	
.40	

As you can see the question, which was posed at, the beginning of this section can be asked for any value of **p**. It should be noted that many times investigators are in fact interested in knowing if it is likely that the true value of **p** is **0.50**. This is simply because a **p** value of **0.50** implies that the population is evenly split between **Yes's** and **No's**. In many cases this indicates that there is nothing of interest happening as a purely random situation would yield **50%** successes and **50%** failures. Thus an investigation of **p = 0.50** is often done to see if there is anything other than random variation.

In this section you have seen the beginnings of the tool called hypothesis testing. This tool will be looked at in much greater detail in Chapters 10, 11, and 12.

Section 7.9 Investigative Exercises

1. Investigate the behavior of the binomial distribution for $p = 0.3$ by varying the value of **n**.

a) Generate and plot the binomial distribution for $p = 0.3$ and $n = 10$.

b) Repeat part a) for $p = 0.3$ but let $n = 15, 20, 25, 30, 50, 100$.

c) Describe the shape of the distribution as **n** increases.

---

2. Investigate the behavior of the binomial distribution for $n = 30$ and $p = 0.05, 0.20, 0.35, 0.50, 0.65, 0.80, 0.95$

a) In each case, generate the probability distribution and then look at the graph.

b) Describe what happens to the shape of the distribution as **p** varies from a small value such as **0.05** to a larger value such as **0.95**.

c) Examine the graphs for $p = 0.20$ and $p = 0.80$. What do you notice? Now look at the graphs for $p = 0.05$ and $p = 0.95$. Do you see the same type of relationship?   Make a general statement, which reflects what you have noticed.

d) What is special about the graph for $p = 0.5$?   Why should you have expected it to look this way?

---

3. Continue the investigation of the variable *Problem*, which was started in the chapter. Are there any differences among the percentage of respondents experiencing recession-related problems for the various industries?

4. Summarize what you have found out about the variable *Problem*. This should include an estimate of **p** for the entire population, estimates of **p** for the population broken down by size and by industry as well as some discussion of any differences you find.

---

5. Investigate the variable *Understands*. Your analysis should follow the same format that was used for the variable *Problem*.

---

6. Investigate the variable *Concerned*.

---

7. Investigate the variable *Call*.

---

8. Is it likely that the true proportion of businesses that feel the bank does not understand their business is **0.20? 0.30? 0.40? 0.50? 0.60? 0.70?**

---

9. Is it likely that the true percentage of companies in this New England City who are concerned that their note may be recalled is **0.50?**

---

10. What recommendations would you make to the Chamber of Commerce?

Chapter 8   "Tissue Strength"

A Study of the Normal Distribution

Section 8.1 Overview

Statistical Objectives: After reading this chapter and doing the exercises a student will:
- Know the meaning of a **z**-value from a standard normal distribution.
- Know how to calculate left and right tail probabilities from a normal distribution.
- Know how changing the mean and standard deviation of a normal distribution affects the left and right tail probabilities.
- Know how changing the mean and standard deviation of a normal distribution affects the position and shape of the distribution.
- Know how to compare real data to a theoretical normal distribution.

Section 8.2 Problem Statement

Large companies solicit consumer complaints to try to correct problems in the manufacturing process that contribute to the number of complaints. Customers feel that the company is listening to them, and the company can try to correct problems before they lose customers to competitors.

In Chapter 4 we learned about some of the customer complaints that a company that manufactures tissues can get. One of the categories of complaints that made up a large percentage of the total was Dispensing. In that category, Sheets Tear on Removal was a significant factor.

The management of the tissue company has decided to address this problem. They know that tensile strength is the factor that determines when a tissue will tear, and have decided that to solve the problem they will have to investigate the tensile strength of the tissues.

As part of the Quality Control program at the company, facial tissue has certain product specifications, that is, criteria that must be met, in order for the product to be acceptable to consumers. One of the characteristics that is specified is tensile strength.

The group has decided to look at the current levels of tissue strength. They know the target values for the process, and the parameters that should be met, and have decided to check to see if the process is meeting the current specifications. If it is not, then changes will need to be made to see that it does. If it is, then perhaps the process specifications will need to be changed. They will collect data on two variables:

> **Machine Direction (MD) Strength:** This is the strength in the direction that the machine pulls on the tissue during manufacture. It has to be high enough that the tissues do not break, causing machine down time.
> **Cross Direction (CD) Strength:** This is the strength in the direction that the tissue is pulled out of the box. It is the variable that would control Sheets Tear.

One of the other variables that needs to be considered is some measure of total strength, which is related to both the dispensing defects and another important tissue parameter, softness. Often strength and softness are tradeoffs. Using both MD and CD Strength, they can calculate the **Geometric Mean Tensile (GMT),** which is equal to the square root of the product of the two variables. This variable is also subject to process specifications.

Samples were taken from tissue produced on a single tissue machine. The samples were taken over three different days and the results were recorded.

Section 8.3 Characteristics of the Data Set

FILENAME:	Ch08Dat.xls	An Excel Workbook
SIZE:	COLUMN	3
	ROWS	200

The first seven lines of the data file are shown in Figure 8.1.

	A	B	C	D	E
1	Day	MDStrength	CDStrength		
2	1	1006	422		
3	1	994	448		
4	1	1032	423		
5	1	875	435		
6	1	1043	445		
7	1	962	464		

Ch08Dat.xls — ◄ ◄ ► ►► \ Data / NormalDist / One ◄

Figure 8.1   Fragment of the Tissue Strength Data File

Notes on the Data file:

1. The variable *Day* keeps track of the day on which the sample was taken and goes from 1 to 3.

2. The variable *MDStrength* measures Machine-directional Strength and is measured in lb./ream.

3. The variable *CDStrength* measures Cross-directional Strength and is measured in lb./ream.

Open the data set named **Ch08Dat.xls** from your data disk. Note that Procedure 3.1, on page 33, contains detail instruction about how to open an Excel workbook.

Click the **Open** *button or use the* **File | Open** *command (* ALT + F *,* O *) to open the file.*

Section 8.4 Normal Distribution in Excel

In order to determine whether the tissue being produced meets the specifications of the manufacturing process is necessary to see what the theoretical distribution of tissue strength looks like. Very often, measurement data is assumed to be normally distributed. This is very often true not only because of the nature of the data itself, but also because of the nature of physical measurement. If a single item is measured for some characteristic such as length, the measurements will vary due to *measurement error*. Most measurements will center on the true value, but some will be higher and some lower. The further a measurement gets from the true value, the less likely it is to occur. This is in fact characteristic of the normal distribution.

According to the specifications, the **MDStrength** is supposed to be normally distributed with a mean of **1000** and a standard deviation of **50** lb./ream. The **CDStrength** should be normally distributed with a mean of **400** and a standard deviation of **25** lb./ream.

Note:
The **Normal Probabilities** *model is included in the* **Ch08Dat.xls** *file. You may wish to check the formulas defined in* **C9:C11** *and* **C15**.

This model can also be generated by the **NormalProbability** *macro.*

Ch08Dat.xls			
	A	B	C	D
1	Normal Probabilities			
2	Parameters			
3	Mean	μ	100	
4	Standard Deviation	σ	20	
5	Limits			
6	Lower Limit	a	70	
7	Upper Limit	b	110	
8	Interval Probabilities			
9	Left-Interval Probability	P(X ≤ a)	0.0668	
10	Right-Interval Probability	P(X > b)	0.3085	
11	Mid-Interval Probability	P(a < X ≤ b)	0.6247	
12	Cumulative Probability			
13	Left-Interval Probability	p = P(X ≤ x)	0.9	
14	Percentile			
15	p^{th} Percentile	x	125.63	

Data \ **NormalDist** \ OneNormal

Figure 8.2   Calculating Normal Probabilities and Percentiles in Excel

Excel provides all functions necessary to calculate normal probabilities and percentiles. The **NormalDist** sheet (Figure 8.2) contains a model that will allow you to answer all typical questions regarding normal probabilities.

Suppose that you want to know the probability that a tissue chosen at random will have **MD Strength** that is **less than or equal to** a specific value **(a=1,000)**; P(X ≤ a) = ?.

Procedure 8.1   Calculating the Left-Interval Normal Probability

Task Description	Mouse/Keyboard
❶ Switch to **NormalDist** sheet and go to cell **C3**.	[IK][◄][►][►I] Data NormalDist [F5] C3 [Enter]
❷ Enter the **MDStrength** mean of **1000** in **C3**, the standard deviation of **50** in **C4**, and go down to **C6**.	1000 [↵] 50 [↵][↵]
❸ Enter the **MDStrength** value of **1000** in **C6** and go down to **C9**.	1000 [↵][↵][↵]

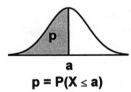

p = P(X ≤ a)

The result is returned in cell **C9** by means of the following Excel function:

$$\overset{\text{a} \quad\text{m}\quad\text{ s}}{\text{=NormDist(C6,C3,C4,True)}}$$

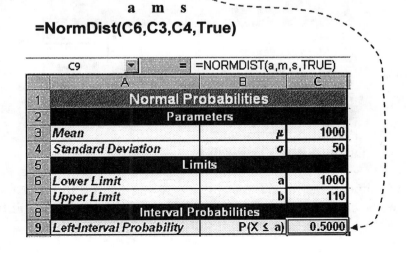

C9	▼	= =NORMDIST(a,m,s,TRUE)

	A	B	C
1	Normal Probabilities		
2	Parameters		
3	*Mean*	μ	1000
4	*Standard Deviation*	σ	50
5	Limits		
6	*Lower Limit*	a	1000
7	*Upper Limit*	b	110
8	Interval Probabilities		
9	*Left-Interval Probability*	P(X ≤ a)	0.5000

Note:
The cells **C3**, **C4** *and* **C6** *are named as* **m**, **s**, *and* **a**, *respectively.*

Note:
The last argument of the **NormDist()** *function takes on either* **True** *or* **False**. **True** *produces a value of the cumulative probability,* **False** *returns a value of the density function.*

Given $\mu = 1000$, and $a = 1000$, the result $P(X \le a) = 0.5$ is obvious. This value corresponds to the Standard Normal probability for $z = 0$. In order to see it in Excel, enter in any free cell this function: **=NormSDist(0)**.

To determine the probability that a tissue chosen at random will have *MDStrength* that is **greater than** a specific value (**b = 1,100**), $P(X > b) = ?$, follow these steps.

Procedure 8.2   Calculating the Right-Interval Normal Probability

	Task Description	Mouse/Keyboard
❶	Assuming the mean and standard deviation of are already defined in **C3** and **C4** (see Procedure 8.1), ◆ go to cell **C7**, ◆ type **1100** there, and move down to cell **C10**.	❶ [F5] C7 [Enter] ❷ 1100 [↓][↓][↓]

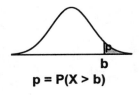

$$p = P(X > b)$$

Note:
The cells **C3**, **C4** *and* **C7** *are named as* **m**, **s**, *and* **b**, *respectively.*

Note:
In order to find the probability that a tissue will have a MD Strength **greater than** *a given value* (**b**), *simply find the corresponding cumulative probability,* $P(X \le b)$, *and subtract it from* **1**.

This time, the result is returned in cell **C10** by means of the following formula:

$$\overset{\text{b m s}}{\text{=1-NormDist(C7,C3,C4,True)}}$$

C10	▼	=	=1-NORMDIST(b,m,s,TRUE)

	A	B	C
1	**Normal Probabilities**		
2	**Parameters**		
3	Mean	μ	1000
4	Standard Deviation	σ	50
5	**Limits**		
6	Lower Limit	a	1000
7	Upper Limit	b	1100
8	**Interval Probabilities**		
9	Left-Interval Probability	$P(X \le a)$	0.5000
10	Right-Interval Probability	$P(X > b)$	0.0228

The formula is based on the fact that $P(X \le b) + P(X > b) = 1$. Now, check the corresponding value for the Standard Normal probability for **z = 2**. In order to check it in Excel, enter in any free cell this function: **=1-NormSDist(2)**. They are the same!

Is this an amazing coincidence? Not really. Remember that a **z**-value (or **z**-score) measures the number of standard deviations (σ) that an observation is from its mean (μ). In this case, **1100** is **100** units (or **2** standard deviations, **2x50**) away from **1000** (μ).

The probabilities **P(X ≤ a)** and **P(X > b)** are examples of left-interval (or left-tail) and right-interval (or right-tail) probabilities. They are the same as the areas under the normal curve from –∞ to **a** and from **b** to +∞, respectively. Notice that the model (Figure 8.2) also shows (in cell **C11**) the probability that an observation will fall between **a** and **b**.

Procedure 8.3   Calculating the Mid-Interval Normal Probability

	Task Description	Mouse/Keyboard
❶	Enter the mean in **C3**, standard deviation in **C4** and lower limit in **C6**.	**Procedure 8.1**
❷	Enter the upper limit in **C7** and move to cell **C11**.	**Procedure 8.2**

The result is returned in cell **C11**. - - - - - - - - - - -

	C11	▼	=	=1-(C9+C10)
	A		**B**	**C**
1	**Normal Probabilities**			
2	**Parameters**			
3	*Mean*		μ	1000
4	*Standard Deviation*		σ	50
5	**Limits**			
6	*Lower Limit*		a	1000
7	*Upper Limit*		b	1100
8	**Interval Probabilities**			
9	*Left-Interval Probability*		P(X ≤ a)	0.5000
10	*Right-Interval Probability*		P(X > b)	0.0228
11	*Mid-Interval Probability*		P(a < X ≤ b)	0.4772

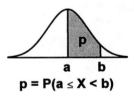

p = P(a ≤ X < b)

Of course, if we know the left-interval probability, $P(X \leq a)$, and the right- interval probability, $P(X > b)$, the mid-interval probability will be:

$$P(a < X \leq b) = 1 - [P(X \leq a) + P(X > b)]$$

or, using the language of Excel:

=1-(C9+C10)

For the *MDStrength* variable, this is the probability that a randomly selected tissue will have a **MD Strength** between **1000** and **1100**.

Exercise 1. What is the probability that a randomly selected tissue will have a **MD Strength** that is equal to or less than **850**? Verify that this value is the same as the z-score for **z = -3**.

Exercise 2. What is the probability that a randomly chosen tissue will have a **MD Strength** between **850** and **1150**?

Exercise 3. What percentage of the tissues should have an **MD Strength** that is greater than **1150**? Verify that this value is the same as the z-score for **z = 3**.

Recall that 25th, 50th, and 75th percentiles are referred to as the first quartile, the second quartile (or median), and the third quartile, respectively.

In many decision situations, we are interested in some "critical" observations (**x**) that are associated with given probabilities (**p**). For example, we might want to know what critical value (**x**) will guarantee a **p=95%** probability of the tissue strength not exceeding that critical value.

Procedure 8.4   Calculating a Normal Percentile

	Task Description	Mouse/Keyboard
❶	Enter the mean in **C3**, standard deviation in **C4**.	**Procedure 8.1**
❷	◆ Go to cell **C13**, ◆ Type **0.95** there, and move down to cell **C15**.	[F5] C13 [Enter] 0.95 [↓][↓]

In this case, we want to know the **95**[th] percentile of this random variable. The probability (**p**) is defined in cell **C13**, and the **p**[th] percentile is returned in cell **C15**.

x ▼	= =NORMINV(p,m,s)	
	A	**B**	**C**
1	Normal Probabilities		
2	Parameters		
3	Mean	μ	1000
4	Standard Deviation	σ	50
5	Limits		
6	Lower Limit	a	1000
7	Upper Limit	b	1100
8	Interval Probabilities		
9	Left-Interval Probability	P(X ≤ a)	0.5000
10	Right-Interval Probability	P(X > b)	0.0228
11	Mid-Interval Probability	P(a < X ≤ b)	0.4772
12	Cumulative Probability		
13	Left-Interval Probability	p = P(X ≤ x)	0.95
14	Percentile		
15	p[th] Percentile	x	1082.24

This is an "inverse" operation when compared to calculating the probability. Therefore the formula in cell **C15** is based on the inverse normal probability distribution function:

$$\begin{matrix} p & m & s \\ =InvNorm(C13,C3,C4) \end{matrix}$$

As you can see, the **95**[th] percentile for the *MDStrength* variable is **1082.24**. You may wish to check, if **P(X ≤ 1082.24) = 0.95**.

Only a few Excel functions have their inverse companions. Nevertheless, in many cases, we can solve "inverse" problems by using the **Tools | Goal Seek** command. Since we already know the **95**th percentile returned by the **NormInv( )** function, it might be interesting to find out if the **Goal Seek** command will lead us to the same solution. The following procedure shows how to do it.

Note:
The cells **C3**, **C4**, **C13** *and* **C15** *are named as* **m**, **s**, **p**, *and* **x** *respectively.*

Note:
You can "play" your own **Goal Seek** *by doing an "intelligent" trial and error experiments. For example, keep trying different values in* **C6** *until* **C9** *is close enough to* **0.95**. *When done,* **C6** *will become the* **95***th percentile.*

Procedure 8.5   Using Goal Seek to Compute a Percentile

	Task Description	Mouse/Keyboard
❶	Go to cell **C9**.	❶ [F5] C9 [Enter]

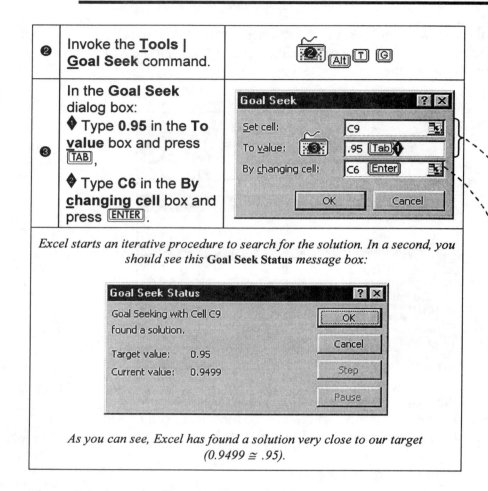

| ❷ | Invoke the **Tools \| Goal Seek** command. | |
| ❸ | In the **Goal Seek** dialog box:
◆ Type **0.95** in the **To value** box and press TAB,

◆ Type **C6** in the **By changing cell** box and press ENTER. | **Goal Seek** [?][X]
Set cell: C9
To value: .95 [Tab]
By changing cell: C6 [Enter]
[OK] [Cancel] |

Excel starts an iterative procedure to search for the solution. In a second, you should see this **Goal Seek Status** *message box:*

Goal Seek Status [?][X]

Goal Seeking with Cell C9
found a solution.

Target value: 0.95
Current value: 0.9499

[OK]
[Cancel]
[Step]
[Pause]

As you can see, Excel has found a solution very close to our target
(0.9499 ≅ .95).

Figure 8.3 show the **Goal Seek** outcome.

	A	B	C
1	**Normal Probabilities**		
2	**Parameters**		
3	*Mean*	μ	1000
4	*Standard Deviation*	σ	50
5	**Limits**		
6	*Lower Limit*	a	1082.2
7	*Upper Limit*	b	1100
8	**Interval Probabilities**		
9	*Left-Interval Probability*	P(X ≤ a)	0.9499

Figure 8.3   Goal Seek Solution (C6) for the
95th Percentile (C9)

Exercise 4. What are the first, the second, and the third quartiles of the *MDStrength* variable?

Exercise 5. For the *MDStrength* variable, find a distance (**d**) from the mean (μ) for which the probability that a randomly selected tissue will have a **MD Strength** between μ - **d** and μ + **d** is **0.95**.

Hint: The best way to start solving this problem is to first visualize it. Figure 8.4 shows the interval (μ - **d**, μ + **d**) and the area above it. The value of the mean is known (μ = **1000**). Thus, if you knew one of the limits, for example, **x** = μ + **d**, then you could determine the distance as **d** = **x** - μ. Notice that the (μ - **d**, μ + **d**) interval is symmetric about the mean (μ) and the area above is **0.95**. Moreover, from Figure 8.4, you can deduce that the limit **x** = μ + **d** is a what percentile?

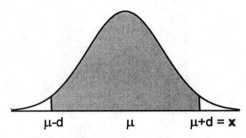

Figure 8.4 d: P(μ-d < X ≤ μ+d) = p

Section 8.5   Plotting the Normal Distribution

In order to plot the normal distribution of the **MD** tissue **Strength**, you need two columns of data: the **X**-axis column representing the values of the *MDStrength* variable and the **Y**-axis column containing the corresponding values of the density function.

When you were setting the variable limits for the Binomial Distribution in Chapter 7 the choices for the beginning, ending and step values for the table were obvious. To decide on the same values for a normal density function requires a bit more thinking.

For a normal distribution, **99.73%** of the area under the probability (density) curve lies within **3** standard deviations of the mean. Just to make sure that you have practically all of the area represented start your **X**-axis range **3** standard deviations below the mean and end **3** standard deviations above the mean.

Note:
Remember, for a continuous random variable the probability distribution is referred to as a density function.

Note:
In Excel, to generate a value of the density function, f(x), you use the same function as for the cumulative distribution, however, the last argument of the function is to be set to False:
=NormDist(x,μ,σ,False)

The **OneNormalPlot** sheet contains a line chart of the normal distribution that is set up based on this $\mu\pm3\sigma$ rule (Figure 8.5). The chart is made of **101** connected points, creating an illusion of a smooth curve. A printed copy of the chart looks even better.

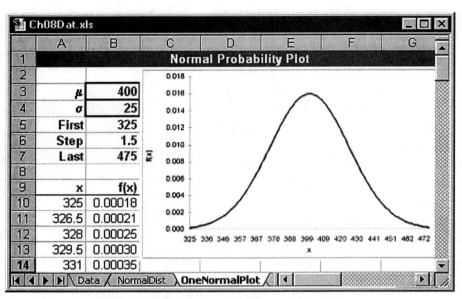

Figure 8.5   Normal Curve for μ=400 and σ=25

In order to plot another normal distribution, you only need to redefine μ and σ in cells **B3** and **B4**, respectively. Check the formulas in **B5**, **B6**, and **B7**. They specify the **First**, **Step**, and **Last** value of **x**. The first and last values are defined as μ-3σ and μ+3σ, respectively. Range **A10:A110** defines the X-axis values, utilizing the first value taken from cell **B5** and increasing recursively every subsequent value by the increment taken from **B6**. Range **B10:B110** contains the corresponding values for the normal density function.

To see the plot for the **MDStrength** variable, enter **1000** into cell **B3** and **50** into cell **B4**. You will see that the shape of the curve has not changed. Only the X-axis labels reveal the new range of the variable. In order to see a difference between two normal distributions, you need to plot them on the same chart (utilizing a common range of **X**).

Section 8.6 Changing the Normal Distribution

There are two parameters that determine the way a normal curve looks, the mean and the variance or standard deviation. The **mean** determines where on the number line the curve is **centered** and the **standard deviation** determines how the curve **spreads out** around the mean.

If you change the mean, the normal curve will be shifted horizontally. Increasing the mean will shift the curve to the right, whereas decreasing the mean will move the curve to the left. The mean does not have any impact on the shape of the curve. This is where the standard deviation reveals its nature. Increasing the standard deviation will flatten the curve (the probability "mass" will get more dispersed). On the other hand, decreasing the standard deviation will slenderize the curve (the probability "mass" will get more concentrated around the mean).

In order to see the real impact of the change it is best to view the two normal curves on the same graph. Switch to sheet **TwoNormalPlots** and define the probability densities for the normal curve with $\mu = 1000$ (in **B4**) and $\sigma = 50$ (in **B5**) and for $\mu = 975$ (in **C4**) and $\sigma = 50$ (in **C5**). Figure 8.6 shows the outcome.

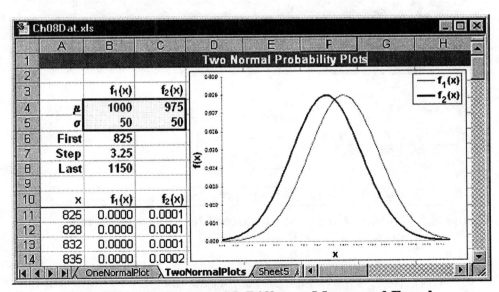

Figure 8.6   Normal Curves with Different Means and Equal Standard Deviations

As you can see, the new distribution ($\mu = 975$ and $\sigma = 50$) got shifted but it has the same shape as the old one ($\mu = 1000$ and $\sigma = 50$).

Note:
To find **P(X≤850)** *and*
P(X>1150) *for the new*
distribution, switch to
sheet **NormalDist** *and*
then redefine μ **(C4)**
and, if necessary, also
a **(C7)** *and* b **(C8)**.

Having done Exercises 1 and 3, you know that the probability that *MDStrength* will fall below **850** or above **1150** is equal **2*0.135% = 0.27%**. For the new distribution ($\mu = 975$ and $\sigma = 50$) this probability is **0.621%+0.023% = 0.644%**. As you can see the percentage of **MD Strengths** below 850 increases from **0.135%** to **0.621%** while the percentage above 1150 decreases from **0.135%** to **0.023%**. It would seem that decreasing the *MDStrength* specifications by **25** lb./ream will not have a negative effect on the sheets tear problem and may in fact help the softness problem.

Exercise 6. Plot two normal curves for the same mean but different standard deviations.

Figure 8.7 shows such curves for $\mu = 1000$ (defined in cells **B4** and **C4**) and for $\sigma_1 = 50$ (**B5**) and $\sigma_2 = 100$ (**C5**). You can make the changes by yourself right on the **TwoNormalPlots** sheet.

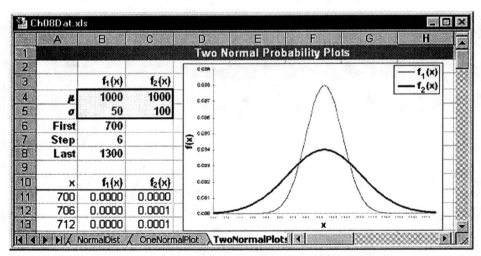

Figure 8.7   Normal Curves With Equal Means and Different Standard Deviations

Exercise 7. Find out the probability that the **MD Strength** will fall below **850** or above **1150** for the new distribution ($\mu = 1000$ and $\sigma = 100$). Compare the results with the old ones.

Hint:

Instead of "recycling" the **Normal Probability** model on the sheet **NormalDist**, run the **NormalProbability** macro.

Procedure 8.6   Using the NormalProbability Macro

	Task Description	Mouse/Keyboard
❶	Execute the **Tools** | **Macro** | **Macros** | **NormalProbability** | **Run** command.	☰❶ [Alt] [T] [M] [M] [Enter]
❷	Define the **Normal** distribution parameters: ◆ Type **1000** for μ and press [TAB], ◆ Type **100** for σ and press [TAB] again.	**Normal Distribution Parameters** Mean - μ Standard Deviation - σ [1000 [Tab]❶] [100 [Tab]❷]
❷	Define the interval limits (**a**, **b**):parameters: ◆ Type **850** for **a** and press [TAB], ◆ Type **1150** for **b** and press [ENTER].	**Probability Input - Interval Limits** a - Left Limit b - Right Limit [850 [Tab]❶] [1150 [Enter]❷]

*The macro calculates the interval probabilities as shown in Figure 8.8. The mid-interval probability box, **p = P(a < X <= b)**, shows the answer to your exercise question.*

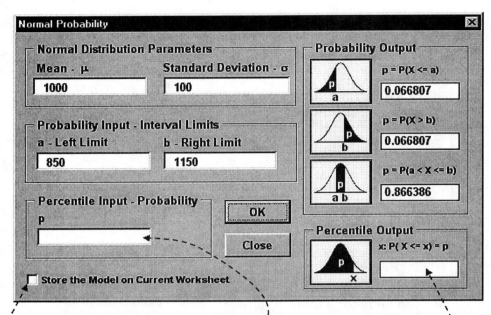

Figure 8.8   Interval (Normal) Probabilities Computed by the NormalProbability Macro

Note:

If you entered a value for **p**, the macro would also produce the **p**[th] percentile. If you checked the **Store the Model on Current Worksheet** box **on**, the macro would replicate its output on the current sheet, starting from the **A1** cell (make sure that there is nothing important in the **A1:C15** range.).

Section 8.7   Empirical Rules for Testing Normality

The specifications assume that the MD strength measurements are normally distributed. If they are not, then even if the mean and standard deviation of the data match the specifications, the probabilities of having defective tissues will not be what you expected! Although there are more formal ways to test to see if data are normally distributed, you can use what you know about the normal distribution to make an empirical (approximate) check.

Anytime that you are examining data you should summarize the data both descriptively and graphically. These methods give you insights that you cannot get from just looking at numbers, and they enable you to understand, intuitively, the results of more sophisticated statistical tests.

Since you are interested in knowing whether or not the process data conforms to the specifications, you want to know if the mean and the standard deviation are correct and whether the distribution is normally distributed. In later chapters you will learn exact statistical tests to answer these questions, but right now you can get a good idea of whether or not they are correct.

The first step then, is to get a set of summary statistics for the *MDStrength* variable. In Chapter 6, you learned how to calculate the sample mean and standard deviation via the **Tools | Data Analysis | Descriptive Statistics** command. You have now a perfect opportunity to explore this command again.

Procedure 8.7   Generating *MDStrength* Summary Measures

	Task Description	Mouse/Keyboard
❶	Switch to the **Data** sheet.	13 712 0.0000 0.000 ◄◄ ◄ ► ►► \ Data \ NormalDist \
❷	Name the data columns using the names stored in the top row.	**Procedure 5.1** (page 100)
❸	Invoke the **Tools	Data Analysis	Descriptive Statistics** command.	Tools Data Analysis... Analysis Tools Anova: Single Factor Anova: Two-Factor With Replication Anova: Two-Factor Without Replication Correlation Covariance Descriptive Statistics Exponential Smoothing F-Test Two-Sample for Variances
	Excel opens the **Descriptive Statistic** *dialog box.*	

Note:
In order to name the columns of data in the data set using the top row labels, switch to the **Data** *sheet, press* CTRL + HOME, *then hold down* SHIFT *and press* END, →, END, ↓. *Having the data set selected, invoke the* **Insert | Name | Create | OK** *command.*

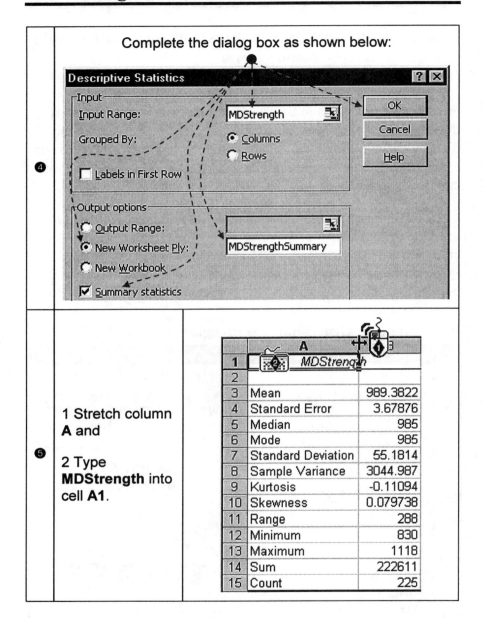

Complete the dialog box as shown below:

❹

❺ 1 Stretch column **A** and

2 Type **MDStrength** into cell **A1**.

From this summary statistics report, you see that the sample mean of the *MDStrength* variable is approximately **989** and the standard deviation is about **55**. While these values are not exactly the **1000** and **50** of the specifications they would appear to be close. Comparing the median, which is approximately **985** to the mean, it appears that the distribution of *MDStrength* is symmetric, which is one of the important characteristics of the normal distribution.

Exercise 8. In the summary statistics report, is there another measure that would suggest an approximate symmetry of the *MDStrength* distribution?

The next step is to view the data graphically to see if the assumption of normality even makes sense. To do this, you will need to create a frequency histogram for the variable *MDStrength*.

Rather than relying on the default class intervals (bins), you will create the following bin range:

(up to 800],(800-850],(850-900],(950-1000], (1000-1050],(1050-1100],(above 1100)

Procedure 8.8   Generating a Frequency Histogram Based on a Custom Bin Range

Task Description	Mouse/Keyboard
❶ On the **MDStrengthSummary** sheet, ◆ enter label **Bin** in **D1** and the series **800**, **850**, ..., **1100** below, ◆ Right-align cell **D1**, and ◆ **Bold** cell **D1**.	
❷ Name the class interval series of numbers (range **D2:D8**) as **Bin**.	**Procedure 5.5** (page 107)
❸ Invoke the **Tools** | **Data Analysis** | **Histogram** command.	

Note:
Procedure 5.12 shows how to enter an arithmetic series in a column range.

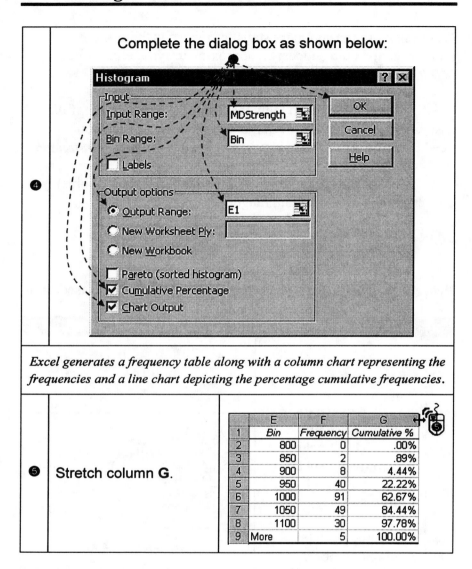

Complete the dialog box as shown below:

Excel generates a frequency table along with a column chart representing the frequencies and a line chart depicting the percentage cumulative frequencies.

❺ **Stretch column G.**

	E	F	G
1	Bin	Frequency	Cumulative %
2	800	0	.00%
3	850	2	.89%
4	900	8	4.44%
5	950	40	22.22%
6	1000	91	62.67%
7	1050	49	84.44%
8	1100	30	97.78%
9	More	5	100.00%

Note:
Procedure 4.2-4.6 shows how to edit and format a chart.

Figure 8.9 shows the chart. Notice that the chart was moved and edited. Some properties of the chart's objects have been changed or disabled.

Exercise 9. Examine the histogram you just created carefully. Do the data appear to be normally distributed?

The histogram shows a distribution that is symmetric and mound shaped. Thus, it would appear that the normal distribution is a reasonable assumption.

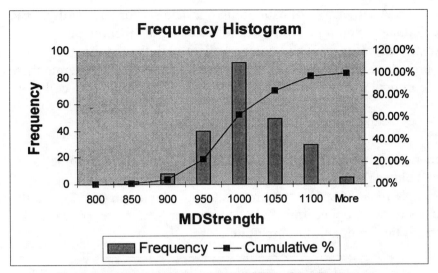

**Figure 8.9   Frequency and Cumulative Frequency
Distributions of *MDStrength***

Since interpreting graphical displays of data is largely
subjective, it is a good idea to use another method in conjunction
with the graph. The Empirical Rule can be used to determine
whether the percent of data falling within **1, 2** and **3** standard
deviations of the mean corresponds to what would be expected
from the normal distribution. Looking at the cumulative
percentage frequencies, you can see that the percent of the data that
are within one standard deviation (between **950** and **1050**) is
84.44% - 22.22 = 62.22%. From the Empirical Rule, approximately
68% of the population should be within one standard deviation of
the mean. Thus, the actual data is not too far off, but you still have
to check the percents within two and three standard deviations to
get the whole picture.

Exercise 10. Use the procedure above to determine the percent of
MDStrength data that are within two and three standard deviations
of the mean.

You should have gotten values of **93.34% (97.78%-4.44%)**
for two standard deviations and **99.11% (100.00%-0.89%)** for three
standard deviations. Comparing these to the values of Empirical
Rule, **95%** and more than **99%** you can conclude that the
assumption of normality is not too far off.

Another way of examining our data set for "normality" is to calculate the cumulative probabilities using the theoretical distribution function and compare them with the empirical percentage frequencies, as shown in Figure 8.10. To generate the theoretical probabilities, enter the following function into cell **H2**:

$$=NormDist(E2,\$B\$3,\$B\$7,True)$$

then copy it and paste onto cells **H3:H9**. Notice that the function calculates the left-interval (cumulative) probability, utilizing the sample mean (**$B$3**), the sample standard deviation (**$B$7**), and the class interval upper limit (**E2**). As you can see, the empirical and theoretical probabilities are quite close.

	E	F	G	H
1	Bin	Frequency	Cumulative %	Cum.Prob.
2	800	0	.00%	0.03%
3	850	2	.89%	0.58%
4	900	8	4.44%	5.26%
5	950	40	22.22%	23.77%
6	1000	91	62.67%	57.63%
7	1050	49	84.44%	86.40%
8	1100	30	97.78%	97.75%
9	1150	5	100.00%	99.82%

◄◄ ► ►◄ **MDStrengthStats** ◄ ►

**Figure 8.10 Comparing Empirical and
Theoretical Probabilities**

Exercise 11. Find out what is the average absolute difference between the empirical and theoretical cumulative probabilities. **Hint:** Use the **Abs( )** function to calculate the absolute differences. Then use the **Average( )** function to determine the average. You should obtain a value of approximately **1.24%**.

Exercise 12. Figure 8.10 shows the theoretical cumulative probabilities based on the sample measures (mean and standard deviation). Create an additional column of the cumulative probabilities, but this time use the process specification measures ($\mu = 1000$ and $\sigma = 50$) rather than the sample measures.

Section 8.8 Investigative Exercises

In the following exercises you are asked to use the skills introduced in the previous chapters to extract information from the tissue strength data file. You are provided with space to answer the questions and paste in graphical output from the program. If you do not have access to a printer, you can sketch the graphs on the axes provided.

1. a) Using the frequency table and histogram that you generated in the preceding chapter, what percent of the *MDStrength* data does not meet the critical specifications? (That is, what percent are actually defective?)

b)   How does this compare to the percent defective expected by the product specifications?

c) Do you think that the company should be concerned about the difference? Why or why not?

---

2. According to the product specifications, **CD Strength** is supposed to be normally distributed with a mean of **450** and a standard deviation of **25**.

a) The critical values for **CD Strength** are **480** on the high side and **390** on the low side. **CD Strength** that is too high creates a stiff tissue. Since the **CD** direction is the one in which tissues are pulled from the box, a value of **CD Strength** that is too low can cause sheets to tear on removal from the carton. According to the specifications, what percent of the tissues should have **CD Strengths** that are too high? too low?

b) What would the lower critical value of **CD Strength** have to be if it were required that the probability of not exceeding be **5%**?

3. Create a graph of the theoretical distribution of **CD Strength**.

---

4. Generate a set of descriptive statistics for the variable *CDStrength*.

a) Compare the mean and the median. Do you think that the distribution of *CDStrength* is symmetric?

b) Looking at the **z**-scores of the largest and smallest values of *CDStrength*, do you think there are any outliers?

c) By comparison, do you think that the *CDStrength* variable meets the values for the mean and the standard deviation that are in the product specifications?

d) Looking at the descriptive statistics, do you think that *CDStrength* is normally distributed?

---

5. a) Create a frequency histogram for the variable *CDStrength*. Does it support the assumption of normality?

b) Compare the percent within **1, 2** and **3** standard deviations to the theoretical values. Are the assumptions of normality still reasonable?

---

6. Another strength criteria which is mentioned in the product specifications is the **Geometric Mean Tensile (GMT).** This is not an actual measurement, but is a value calculated from **MD** and **CD Strength**. The **GMT** is found by taking the square root of the product of **MD** and **CD Strength**.

a) Create a new variable *GMT* in your worksheet.

b) Based on all of the criteria do you think that *GMT* is normally distributed? Include reasons for your conclusions.

---

7.   The critical specifications for **GMT** are **575** on the low side and **700** on the high side.

a) What percent of the product is expected to be defective according to the theoretical specifications?

b) What percent of the actual product is defective?

---

8. Prepare a report to management that indicates whether or not the process appears to be running to the product specifications. In this report include any changes that need to be made (in terms of mean and standard deviation) to bring the process back to target values.

Also make any suggestions for changing the target values that you think are appropriate. You cannot change the critical (defective) values, but you might be able to adjust the target values for the mean and standard deviation that would improve the sheets tearing problem. Include all appropriate tables and graphs in the report.

Chapter 9   "Diaper Weight"

A Study of the Central Limit Theorem

Section 9.1 Overview

Statistical Objectives: After reading this chapter and doing the exercises a student will:

- Know what a sampling distribution is and why it is important.
- Know that the sample mean is normally distributed.
- Know what the Central Limit Theorem means.
- Know the effects of sample size on the distribution of the mean.
- Know the effects of the distribution of the original data on the distribution of the sample mean.
- Know what it means for a measurement to be in statistical control.
- Know how to identify from a control chart when a measurement is in or out of statistical control.

Section 9.2 Problem Statement

Most large manufacturing companies use some form of **Statistical Quality Control** (SQC) in the manufacture of their products. One form of SQC that is often used is a **Control Chart**. A control chart looks at statistics from samples of products taken over time. The sample statistic that is often observed is the sample mean.

A large company that manufactures disposable diapers collects data from its machines at random times during the day. One of the variables that is measured during this time is diaper weight. Diaper weight is an important factor in the manufacturing process for two reasons. The first reason is that the material that contributes most to diaper weight is the most expensive component of the diaper. Thus it is reasonable to want to provide enough of this material, but not an excessive amount. The second reason is that the weight of a diaper relates to the consumer's perception of how well that diaper will absorb liquid. Thus it is also reasonable that the amount of the material be adequate to satisfy the consumers.

When diapers are collected they are collected in samples of size 5. The sample number and the individual diaper weights and bulks are recorded. The sample averages are then calculated and plotted on charts. Machine operators use these charts to tell them whether or not the machine is behaving as expected or whether the machine needs adjustment.

Section 9.3 Characteristics of the Data Set

FILENAME: Ch09Dat.xls   An Excel Workbook
SIZE: COLUMN   4
               ROWS   260

A fragment of the data file is shown in Figure 9.1.

	A	B	C	D	E
1	**Sample**	**Diaper**	**Weight**	**Bulk**	
2	1	1	55.87	0.419	
3	1	2	55.35	0.380	
4	1	3	54.50	0.365	
5	1	4	53.97	0.406	
6	1	5	54.29	0.360	
7	2	1	54.85	0.397	

Figure 9.1   A Fragment of the Diaper Data Set

Notes on the Data file:

1. The variable *Sample* keeps track of the sequence in which the samples were taken and goes from 1 to 52.

2. The variable *Diaper* keeps track of the individual diapers within a sample and goes from 1 to 5.

3. The variable *Weight* records the diaper weight in grams.

4. The variable *Bulk* records the diaper bulk in millimeters.

Open the data set named **Ch09Dat.xls** from your data disk. Note that Procedure 3.1, on page 33, contains detail instruction about how to open an Excel workbook.

Click the Open *button or use the* File | Open *command* (ALT+F, O) *to open the file.*

Section 9.4 Finding Sample Means

In order to see what the machine operators plot for each sample you will have to calculate the mean for diaper weight and bulk for each of the 52 samples. Procedure 7.4 on page 185, demonstrates an example of using the **Data | Pivot Table Report** command to calculate proportions. You will use the same command for the sample means.

Note:
To find out all the sample averages, you could also use the Data | Subtotals *command.* PivotTable *is here more convenient as it generates the output in form of a column of the averages that is ready for further processing.*

Procedure 9.1   Generating sample means for subsets of the variable Weight

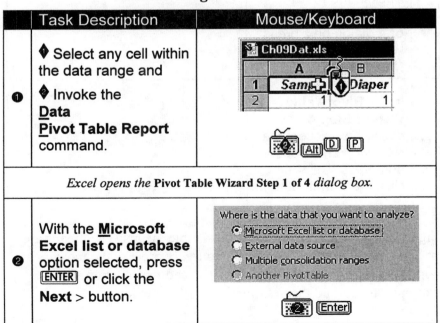

	Task Description	Mouse/Keyboard
❶	◆ Select any cell within the data range and ◆ Invoke the **Data Pivot Table Report** command.	
	Excel opens the **Pivot Table Wizard Step 1 of 4** *dialog box.*	
❷	With the **Microsoft Excel list or database** option selected, press ENTER or click the **Next** > button.	

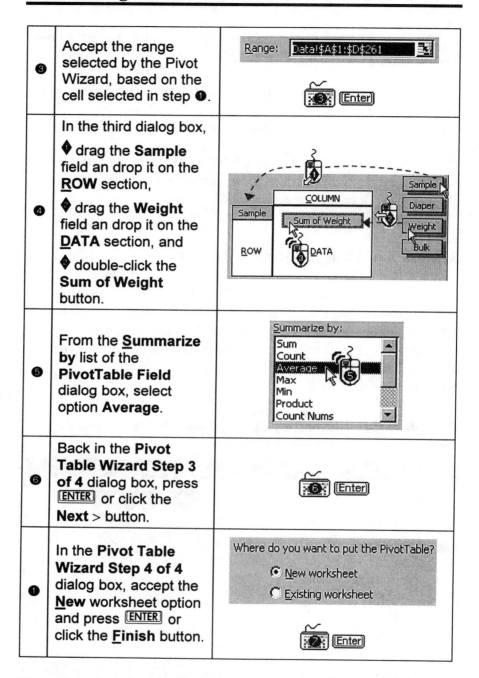

❸	Accept the range selected by the Pivot Wizard, based on the cell selected in step ❶.	
❹	In the third dialog box, ◆ drag the **Sample** field an drop it on the **ROW** section, ◆ drag the **Weight** field an drop it on the **DATA** section, and ◆ double-click the **Sum of Weight** button.	
❺	From the **Summarize by** list of the **PivotTable Field** dialog box, select option **Average**.	
❻	Back in the **Pivot Table Wizard Step 3 of 4** dialog box, press [ENTER] or click the **Next >** button.	
❼	In the **Pivot Table Wizard Step 4 of 4** dialog box, accept the **New** worksheet option and press [ENTER] or click the **Finish** button.	

Figure 9.2 shows a fragment of the sample means. The means are part of the pivot table. To be able to freely process the means, copy the range (**B3:B54**) and paste it to another location, for example, to **C2** and then eliminate the pivot table.

	A	B
1	Average of Weight	
2	Sample	Total
3	1	54.796
4	2	55.134
5	3	54.836
6	4	54.59
7	5	55.268

54	52	55.19
55	Grand Total	54.99480769

Sheet1 / Data / Sheet2 / Sh

Figure 9.2 Sample Means of Weight

In order to facilitate calculations common for both the *Weight* variable and the *MeanWeight* variable, copy the former from the **Data** sheet and paste it next to the latter. Finally, adjust column width and name the sheet tab as **WeightAndMeans**. A fragment of the intended outcome is shown in Figure 9.3. Note that the size of the variables is different. The *MeanWeight* variable contains the averages of 5-element samples taken from the *Weight* variable.

To do this, select range **B3:B54** and click the copy icon. Then move to cell **C2** and click the paste icon. Select columns **A**, **B** and execute the **Edit | Delete** command. Since the columns contain the pivot table, by deleting them the table will also be deleted. Now, label the *Weight* means as *MeanWeight* (type the label in cell **A1**).

	A	B
1	**MeanWeight**	**Weight**
2	54.796	55.87
3	55.134	55.35
4	54.836	54.50
5	54.59	53.97

260		54.65
261		55.29
262		

WeightAndMeans /

Figure 9.3 A fragment of the MeanWeight and Weight variables.

Hint:
To select a range, click its first cell and shift-click the last one.
To select a block of columns, click the header of the first one and shift-click the header of the last one.

Hint:
Procedure 5.10 (page 121) shows how to copy data from one sheet to another.

Exercise 1. Calculate summary statistics, and generate frequency tables and histograms of the variables *MeanWeight* and *Weight*.

Figure 9.5 shows the summary statistics report for the variables and the common class interval setup (**Bin**) for their frequency histograms. The summary measure report was generated using the **Tools | Data Analysis | Descriptive Statistics** command and then it was slightly edited. Figure 9.4 shows the **Descriptive Statistics** dialog box with the **Input Range** including both the variables. Its output is to go to a new worksheet, **WeightAndMeansSummary**.

Hint:
Procedure 6.4 (page 147) shows how to produce the summary measures for two data sets.
Procedures 5.12 and 5.13(pages 124,125) show how to set up a bin range and generate frequency histograms.

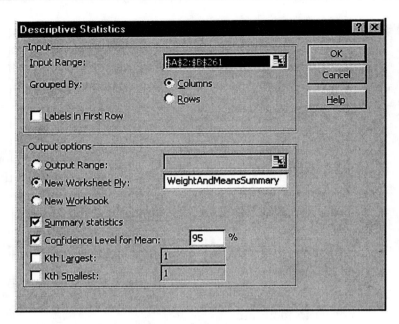

**Figure 9.4 The Setup for the Summary Measures
of the variables MeanWeight and
Weight.**

The **Bin** range for the histograms, was prepared to ensure that it covers all values of the variables. The lower **Bin** limit (53.00) is slightly less than the *Weight* **Min** (53.02) and the upper **Bin** limit (56.75) is a bit greater than the *Weight* **Max** (56.6).

	A	B	C	D
1	**Measures**	*MeanWeight*	*Weight*	*Bin*
2	Mean	54.99480769	54.99480769	53.00
3	Standard Error	0.037107047	0.038744955	53.25
4	Median	55.017	54.95	53.50
5	Mode	55.06	54.89	53.75
6	Standard Deviation	0.267582719	0.624743624	54.00
7	Sample Variance	0.071600511	0.390304596	54.25
8	Kurtosis	-0.758963774	-0.126274086	54.50
9	Skewness	-0.040119224	0.063219354	54.75
10	Range	1.092	3.58	55.00
11	Minimum	54.408	53.02	55.25
12	Maximum	55.5	56.6	55.50
13	Sum	2859.73	14298.65	55.75
14	Count	52	260	56.00
15	Confidence Level(95.0%)	0.074495447	0.076295215	56.25
16				56.50
17				56.75

**Figure 9.5 The Summary Measures and Class Intervals for the
Variables *MeanWeight* and *Weight***

Figure 9.6 shows the frequency distributions **MeanWeightFrq** and **WeightFrq** of the variables *MeanWeight* and *Weight*, respectively, both obtained through the **Frequency( )** function. The following procedure shows how to do it.

	D	E	F
1	*Bin*	*MeanWeightFrq*	*WeightFrq*
2	53.00	0	0
3	53.25	0	1
4	53.50	0	1
5	53.75	0	3
6	54.00	0	6
7	54.25	0	16
8	54.50	1	31
9	54.75	11	38
10	55.00	12	41
11	55.25	19	37
12	55.50	9	28
13	55.75	0	25
14	56.00	0	18
15	56.25	0	8
16	56.50	0	5
17	56.75	0	2

|◀ ◀ ▶ ▶| \ **WeightAndMeansSummary** / Weight

Figure 9.6   Frequency Distributions of
MeanWeight* and *Weight

Procedure 9.2   Generating the Frequency Distributions of
MeanWeight* and *Weight

Task Description	Keyboard
❶ Switch back to sheet **WeightAndMeans** and name: ◆ Name the **MeanWeight (A2:A53)** range as **MeanWeight** and ◆ The **Weight** range **(B2:B261)** as **Weight**.	❶ Ctrl Home Shift End ↓ Alt I N C Enter ❷ Ctrl Home → Shift End ↓ Alt I N C Enter
❷ Go back to sheet **WeightAndMeansSummary** and name the Bin range **(D2:D17)** as **Bin**.	❷ Ctrl Page Up F5 D1 Enter Shift End ↓ Alt I N C Enter

Recall:
You can name a range or cell using the label adjacent to the range by means of the **Insert | Name | Create** *command.*
To select a range filled with data, select the first cell, hold down **SHIFT** *and press the arrow key that points to last cell in the range.*

❸	Select range **E2:E17** for the frequency distribution of *MeanWeight*, type **=Frequency(MeanWeight, Bin)**, and press SHIFT + CTRL + ENTER.	[F5] E2:E17 [Enter] =Frequency(MeanWeight,Bin) [Shift][Ctrl][Enter]
❹	Select range **F2:F17** for the frequency distribution of *Weight*, type **=Frequency(Weight, Bin)**, and press SHIFT + CTRL + ENTER.	[F5] F2:F17 [Enter] =Frequency(Weight,Bin) [Shift][Ctrl][Enter]

Hint:

Procedure 5.13 shows how to produce a two-histogram chart. Here select range **E1:F17**, *invoke the* **Chart Wizard**. *Accept the default settings in* **Step 1**. *Select the* Series *tab in* **Step 2** *and add range* **D2:D17** *to* **Category (X)** *axis labels. In the other steps choose options and settings that will make your chart look like the one shown on the right.*

Figure 9.7 shows the frequency histograms for *Weight* and *MeanWeight* created with a help of the **ChartWizard**.

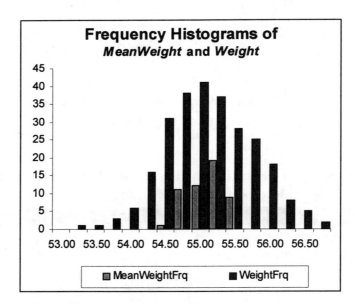

Figure 9.7   Frequency Histograms of
MeanWeight* and *Weight

Exercise 2. Use the histogram and the descriptive statistics to determine whether it is reasonable to believe that the variable *Weight* is normally distributed.

From the shape of the histogram and a comparison of the median and the mean it would appear that diaper weight is normally distributed with a mean of 54.99 gm and a standard deviation of 0.625 gm.

Section 9.5 The Central Limit Theorem

The descriptive statistics for the variable *Weight* shown in Figure 9.5 indicate that the weights of the diapers vary from a minimum of 53.02 to a maximum of 56.60 gm. When the machine operators plot the data on control charts, they look at the *average* of the five diapers in a sample. How does the distribution of the averages compare to the distribution of the individual diaper weights?

From the histogram in Figure 9.7, we see that while *MeanWeight* centers at the same value as *Weight*, it exhibits much less variability. In fact, looking at the descriptive statistics shown in Figure 9.5, the standard deviation of the individual diapers is about 2.2 times that of the sample means!

If you remember the **Central Limit Theorem**, you know that for a random variable X, with mean μ and standard deviation σ, the distribution of the sample means (sampling distribution of $\overline{X}$) is normal with mean μ and standard deviation $\sigma / \sqrt{n}$ (standard error of the mean). This is true *exactly* when the distribution of X is normal and *approximately* when the distribution of X is not normal for large enough values of n. If you consider that the sample size used for *MeanWeight* was n = 5, and that $\sqrt{5} = 2.236...$, then this is *exactly* what you should expect according to the Central Limit Theorem.

Section 9.6 Control Charts in Excel

Section 9.6.1 About Control Charts

In the problem statement, you read that the machine operators plot the sample averages on *Control Charts*. A control chart is a statistical tool for determining when a random variable associated with a manufacturing process is "out of control". "Out of control" means that the measurement of interest is not behaving according to its specifications. The theory of the control chart is based on the Central Limit Theorem, since what is plotted is the sample average!

A control chart is a graph that plots the sample averages in chronological order along the **X** axis. There are three guidelines on a control chart that tell the operator whether the process is in control.

- The **Center Line (CL)** of the chart is drawn at the theoretical or specified value for the mean of the distribution.
- The **Upper Control Limit (UCL)** of the chart is a horizontal line drawn at a value that is three standard errors of the mean

  (called $3\,\sigma_{\bar{x}}$ limits) above the center line.
- The **Lower Control Limit (LCL)** of the chart is a horizontal line drawn at a value that is three standard errors of the mean

  (called $3\,\sigma_{\bar{x}}$ limits) below the center line.

Control limits at $\mu \pm 3\,\sigma_{\bar{x}}$ indicate events that are highly unlikely to occur, if the theoretical distribution of the measurement is correct. Remember from your study of the normal distribution that 99.73% of all values from a normal distribution fall within 3 standard deviations of the mean. Thus, the probability that a sample average will fall outside the control limits is 0.0027 or 27 in 10,000!

As long as the measurements plotted fall within the 3 sigma limits, the process is assumed to be in control and allowed to run. If, however, a point falls outside the limits, either above the **UCL** or below the **LCL**, the process is stopped and the operators search for the cause of the erroneous value:

$$UCL = \mu + 3\frac{\sigma}{\sqrt{n}}$$

$$LCL = \mu - 3\frac{\sigma}{\sqrt{n}}$$

The control limits for a variable can be based on the specifications or derived from sample data. For **Diaper Weight** the target mean is 55 gm. with a target standard deviation of 0.55 gm. Using these values, if you substitute in the formulas, you get:

$$UCL = 55 + 0.738 = 55.738$$

and

$$LCL = 55 - 0.738 = 54.262.$$

When a machine operator looks at the control chart they look to see if any points (*MeanWeight*) fall outside the control limits. If they do, then they conclude that the process average is not in control, that is, it is not at the value indicated by the center line of the chart. They then take action to bring the process back into control.

Section 9.6.2 Creating a Control Chart in Excel

Excel has options available that create control charts for a variety of different situations. To create a control chart based on the sample data (*MeanWeight*) you first need to generate an XY-scatter chart of the averages (Figure 9.8). Again, you will have another opportunity to practice the **Chart Wizard**. The following steps highlight the most important operations.

Procedure 9.3   Generating a Scatter Plot of *MeanWeight*

	Task Description	Mouse/Keyboard
❶	Select range **MeanWeight (A2:A53 on sheet WeightAndMeans)**.	
❷	Invoke the **Chart Wizard**.	

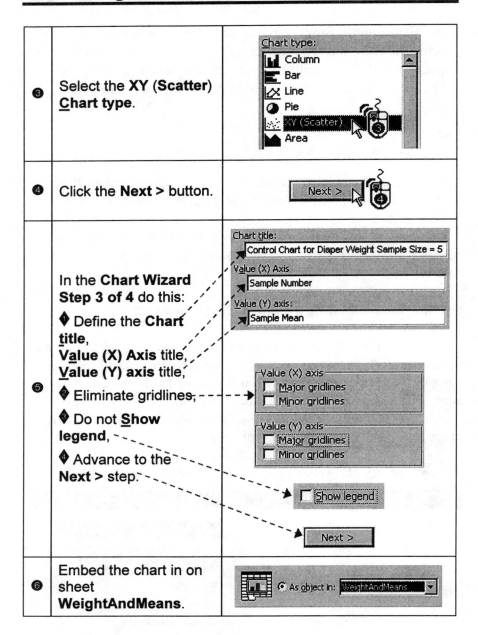

❸	Select the **XY (Scatter)** Chart type.	
❹	Click the **Next >** button.	
❺	In the **Chart Wizard Step 3 of 4** do this: ◆ Define the **Chart title**, **Value (X) Axis** title, **Value (Y) axis** title, ◆ Eliminate gridlines, ◆ Do not **Show legend**, ◆ Advance to the **Next >** step.	
❻	Embed the chart in on sheet **WeightAndMeans**.	

After you have generated the chart, take a few minutes to format it. Do your best to bring the chart as close the one shown in Figure 9.8 as possible.

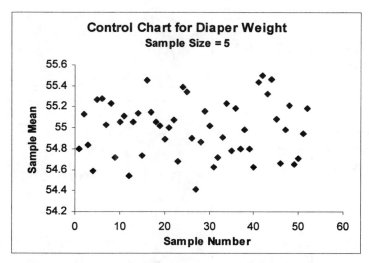

Figure 9.8   Scatter Plot of *MeanWeight*
(Control Chart - Phase 1)

Your next task is to insert the lower and upper 3 sigma limits. The following procedure shows how to do it.

Procedure 9.4   Inserting the 3-Sigma Control Limits

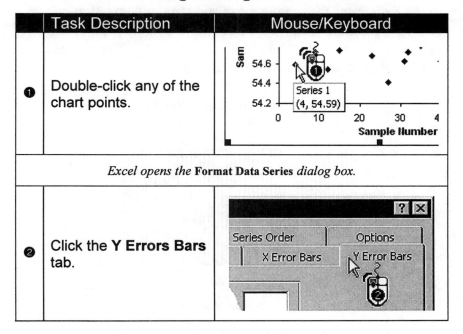

	Task Description	Mouse/Keyboard
❶	Double-click any of the chart points.	
	Excel opens the **Format Data Series** *dialog box.*	
❷	Click the **Y Errors Bars** tab.	

❸	Click the **Display Both** option box.	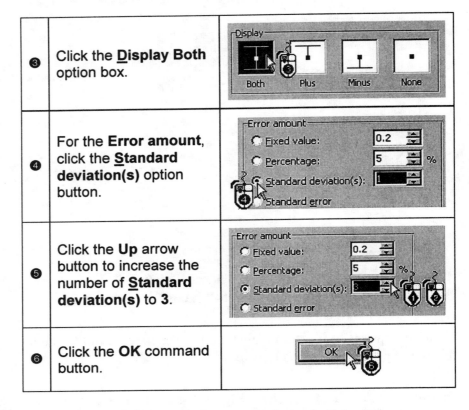
❹	For the **Error amount**, click the **Standard deviation(s)** option button.	
❺	Click the **Up** arrow button to increase the number of **Standard deviation(s)** to **3**.	
❻	Click the **OK** command button.	

Note:
Make sure that the
Standard deviation(s)
option button is turned
on (◉).

Excel uses the sample mean and standard deviation (derived from the **Y** range, here from *MeanWeight*) to plot the Mean ± (3)(Standard Deviation) limits (Figure 9.9). Notice that all the *MeanWeight* points are within the limits.

Note:
Here, the **Y** *range*
Mean *is the mean of*
the **Weight Means** *and*
the **Y** *range* **Standard**
deviation *is the*
standard deviation of
the **Weight Means**.
Thus the limits shown
on this chart are a
good approximation of
*the control limit (**UCL***
and **LCL***) discussed*
earlier in this chapter.
The calculated values
are 55.738 for **UCL** *and*
54.262 for **LCL**.

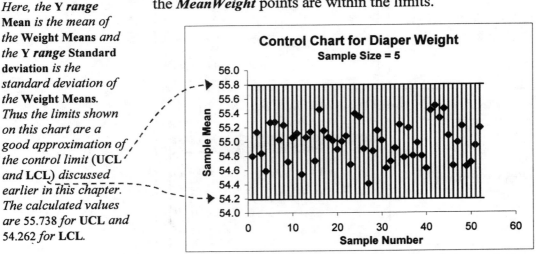

Figure 9.9   Control Chart for Diaper *Weight*

In order to create a control chart with the specification based limits (**LCT** and **UCL**), you need to store the limits in the worksheet and then plot the limits along with the data using the **ChartWizard**. The following procedure shows all necessary steps. Unlike the control limits generated above, in which the means (the *MeanWeight* set) are based on the **5**-element samples, this procedure uses the means derived from **10**-element samples. To distinguish the new set of the means from the old one, we will name as *Mean10Weight*.

Procedure 9.5   Generating the Specification-Based Control Limits for the Variable Weight

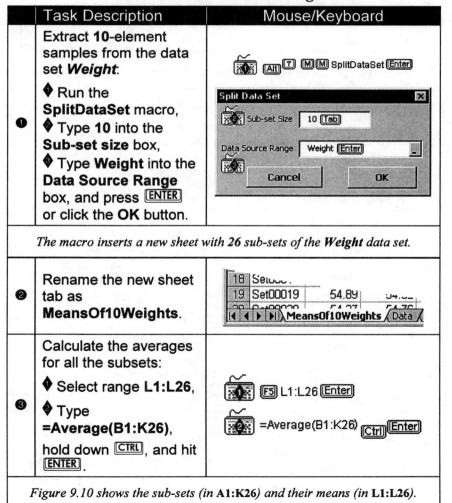

	Task Description	Mouse/Keyboard
❶	Extract **10**-element samples from the data set *Weight*: ◆ Run the **SplitDataSet** macro, ◆ Type **10** into the **Sub-set size** box, ◆ Type **Weight** into the **Data Source Range** box, and press [ENTER] or click the **OK** button.	
	The macro inserts a new sheet with 26 sub-sets of the Weight data set.	
❷	Rename the new sheet tab as **MeansOf10Weights**.	
❸	Calculate the averages for all the subsets: ◆ Select range **L1:L26**, ◆ Type **=Average(B1:K26)**, hold down [CTRL], and hit [ENTER].	
	Figure 9.10 shows the sub-sets (in A1:K26) and their means (in L1:L26).	

Hint:
In order to rename a sheet tab, double-click the tab and enter its new name.

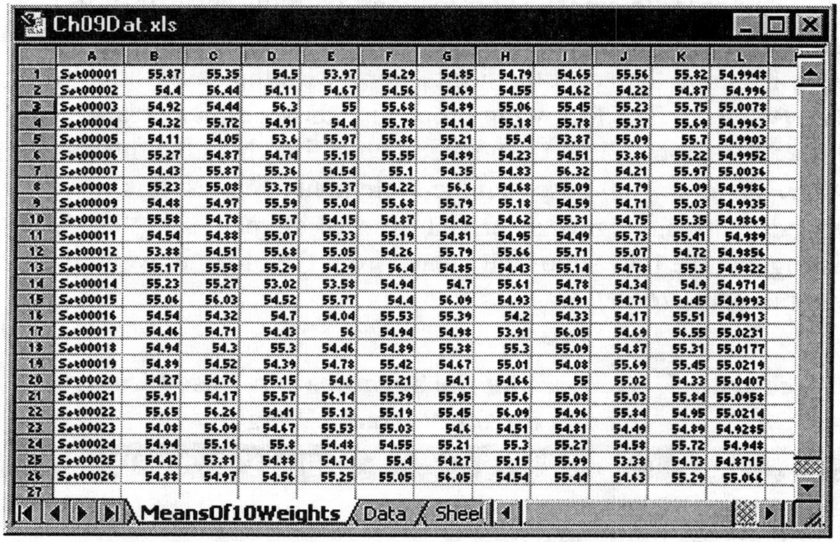

	A	B	C	D	E	F	G	H	I	J	K	L
1	Set00001	55.87	55.35	54.5	53.97	54.29	54.85	54.79	54.65	55.56	55.82	54.9948
2	Set00002	54.4	56.44	54.11	54.67	54.56	54.69	54.55	54.62	54.22	54.87	54.996
3	Set00003	54.92	54.44	56.3	55	55.68	54.89	55.06	55.45	55.23	55.75	55.0078
4	Set00004	54.32	55.72	54.91	54.4	55.78	54.14	55.18	55.78	55.37	55.69	54.9963
5	Set00005	54.11	54.05	53.6	55.97	55.86	55.21	55.4	53.87	55.09	55.7	54.9903
6	Set00006	55.27	54.87	54.74	55.15	55.55	54.89	54.23	54.51	53.86	55.22	54.9952
7	Set00007	54.43	55.87	55.36	54.54	55.1	54.35	54.83	56.32	54.21	55.97	55.0036
8	Set00008	55.23	55.08	53.75	55.37	54.22	56.6	54.68	55.09	54.79	56.09	54.9986
9	Set00009	54.48	54.97	55.59	55.04	55.68	55.79	55.18	54.59	54.71	55.03	54.9935
10	Set00010	55.58	54.78	55.7	54.15	54.87	54.42	54.62	55.31	54.75	55.35	54.9869
11	Set00011	54.54	54.88	55.07	55.33	55.19	54.81	54.95	54.49	55.73	55.41	54.989
12	Set00012	53.88	54.51	55.68	55.05	54.26	55.79	55.66	55.71	55.07	54.72	54.9856
13	Set00013	55.17	55.58	55.29	54.29	56.4	54.85	54.43	55.14	54.78	55.3	54.9822
14	Set00014	55.23	55.27	53.02	53.58	54.94	54.7	55.61	54.78	54.34	54.9	54.9714
15	Set00015	55.06	56.03	54.52	55.77	54.4	56.09	54.93	54.91	54.71	54.45	54.9993
16	Set00016	54.54	54.32	54.7	54.04	55.53	55.39	54.2	54.33	54.17	55.51	54.9913
17	Set00017	54.46	54.71	54.43	56	54.94	54.98	53.91	56.05	54.69	56.55	55.0231
18	Set00018	54.94	54.3	55.3	54.46	54.89	55.38	55.3	55.09	54.87	55.31	55.0177
19	Set00019	54.89	54.52	54.39	54.78	55.42	54.67	55.01	54.08	55.69	55.45	55.0219
20	Set00020	54.27	54.76	55.15	54.6	55.21	54.1	54.66	55	55.02	54.33	55.0407
21	Set00021	55.91	54.17	55.57	56.14	55.39	55.95	55.6	55.08	55.03	55.84	55.0958
22	Set00022	55.65	56.26	54.41	55.13	55.19	55.45	56.09	54.96	55.84	54.95	55.0214
23	Set00023	54.08	56.09	54.67	55.53	55.03	54.6	54.51	54.81	54.49	54.89	54.9285
24	Set00024	54.94	55.16	55.8	54.48	54.55	55.21	55.3	55.27	54.58	55.72	54.948
25	Set00025	54.42	53.81	54.88	54.74	55.4	54.27	55.15	55.99	53.38	54.73	54.8715
26	Set00026	54.88	54.97	54.56	55.25	55.05	56.05	54.54	55.44	54.63	55.29	55.066
27												

Figure 9.10 10-Element Subsets of Weight and Their Means.

❹	Insert a new row before the row number 1: ♦ Select any cell in row **1** and ♦ Invoke the **Insert | Rows** command.	❶ F5 L1 Enter ❷ Alt I R
❺	Enter the following labels into cells L1, M1, N1, and O1: **Mean10Weight** **Mean Weight** **LCL** **UCL**	❺ F5 L1 Enter Mean10Weight → MeanWeight → LCL → UCL Enter
❻	In cell **M2**, enter the **=Average(Weight)** function, to calculate the mean ***Weight***.	❻ F5 M2 Enter =Average(Weight) Enter
❼	In cell **N2**, enter the **=M2-3*StDev(Weight)/Sqrt(10)** formula, to calculate the **Lower Control Limit**.	❼ F5 N2 Enter =M2-3*StDev(Weight)/Sqrt(10) Enter

Hint:
Instead of pressing the function key F5 to jump to a cell or range, you can use the navigation keys or mouse click.

❽	In cell **O2**, enter the =M2+3*StDev(Weight)/Sqrt(10) formula, to calculate the **Upper Control Limit**.	🖩 ❽ [F5] O2 [Enter] =M2+3*StDev(Weight)/Sqrt(10) [Enter]
❾	Replicate the values of **Mean Weight**, **LCL**, and **UCL** throughout range **M3:O27**.	🖩 ❽ [F5] M3:O27 [Enter] =M2:O2 [Ctrl] [Enter]
❿	Insert a column in front of the column **L**.	🖩 ❿ [F5] L1 [Enter] [Alt] [I] [C]
⓫	Label the new column as **Sample Number** in cell **L1** and fill the cells below with numbers **1** through **26**.	🖩 ⓫ [F5] L1 [Enter] Sample Number [↓] 1 [Ctrl] [Enter] [Alt] [E] [I] [S] [Alt] [C] [Alt] [O] 26 [Enter]

Hint:
You can also insert a new column by right-clicking a column header and choosing the **Insert** *option..*

Note:
Procedure 5.12 (page 124) shows detail steps for generating a series of numbers.

Figure 9.11 shows a fragment of the setup for the *Weight* control limits.

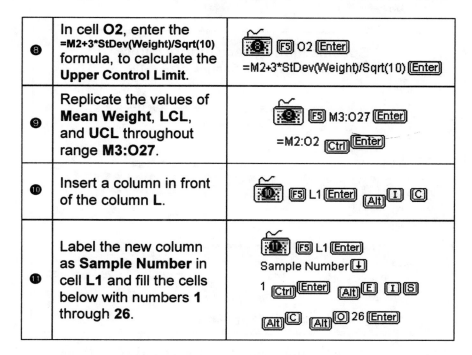

	L	M	N	O	P
1	**Sample Number**	*Mean10Weight*	**Weight Mean**	**LCL**	**UCL**
2	1	54.99	54.99	54.80	55.19
3	2	55.00	54.99	54.80	55.19
4	3	55.01	54.99	54.80	55.19
5	4	55.00	54.99	54.80	55.19
6	5	54.99	54.99	54.80	55.19
7	6	55.00	54.99	54.80	55.19
	7	55.00	54.99	54.80	55.19
24			54.99	54.80	55.19
25	24	54.95	54.99	54.80	55.19
26	25	54.87	54.99	54.80	55.19
27	26	55.07	54.99	54.80	55.19

◄ ◄ ► ►◄ \\ **MeansOf10Weights** / Data / Sheet2 / Sheet3 / Sheet4 /

Figure 9.11 Specification-Based Control Limits for the Variable Weight

Exercise 3. Generate an **XY (Scatter)** chart based on the **L1:P27** data range. Format the chart as shown in Figure 9.12.

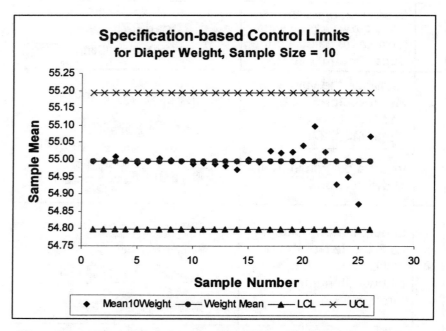

Figure 9.12 Specification-based Control Chart for Diaper Weight

Again, the control chart shows that the process is in control. It is interesting to note that the control limits obtained for the sample size of 10 (LCL = 54.80, UCL = 55.19) are stricter than those for the sample size of 5 (LCL = 54.26, UCL = 55.74). More information provides a higher precision.

Section 9.7 Investigative Exercises

In the following exercises you are asked to use the skills introduced in the previous chapters to extract information from the tissue strength data file. You are provided with space to answer the questions and paste in graphical output from the program. If you do not have access to a printer, you can sketch the graphs on the axes provided.

1. The target values for Diaper Weight are a normal distribution with a mean of 55 and a standard deviation of 0.55. Create plots of the theoretical distributions for *Weight* and *MeanWeight*. How do they compare to the histograms created from the actual data?

_____

2. a) Examine the effects of a change in the sample size on the sampling distribution of the average diaper weight. Create control charts for subgroups of 10, 15 and 20 diapers. Use the data to calculate the values for the Mean (Center Line) and for the UCL and LCL.   Fill out the table below:

Sample Size	Center Line	Standard Error of the Mean	Lower Control Limit	Upper Control Limit
n	(Mu)	($Sigma / \sqrt{n}$)	LCL	UCL
10				
15				
20				

b) What do you notice about the values for the Center Line? Why does this happen?

c) What effect, if any, does a change in the sample size have on the distribution of *MeanWeight*?

d) How do the values for the standard errors compare to the actual values according to the Central Limit Theorem?

3. a) Calculate a set of summary statistics and a histogram for the variable *Bulk*. Does the data appear to be normally distributed?

b) Find the *MeanBulk* for the samples of size 5. (Use the data where the samples are arranged in rows.) Calculate summary statistics and make a histogram for *MeanBulk*. How do they compare to that of *Bulk*? If the target values for *Bulk* are a mean of 0.400 with a standard deviation of 0.02, how do they compare to the values you would expect from the Central Limit Theorem?

c) Vary the sample size for the *MeanBulk* and create control charts to fill in the table below:

Sample Size n	Center Line (Mu)	Standard Error of the Mean (*Sigma* / $\sqrt{n}$)	Lower Control Limit LCL	Upper Control Limit UCL
10				
15				
20				

d) How do the standard errors of the sample means compare to the distribution for the variable *Bulk*? Is this what you would expect? Why or why not?

e) What is the effect of changing the sample size for the variable *MeanBulk*? Is it similar to the effect for *MeanWeight*?

---

4. Look at the control chart for *MeanBulk.*
a) Does the measurement *MeanBulk* appear to be in statistical control? Why or why not?

5. The product specifications for the diapers include a variable *Density* which is found by dividing the *Weight* of a diaper by its *Bulk*.

a) Create the variable *Density,* calculate a set of summary statistics for it and make a histogram of the data. (Use the data from the first columns where the samples are in the columns.) Does this variable appear to be normally distributed?

b) Create control charts for *MeanDensity* using samples of size 5. How does *MeanDensity* compare to *Density*?

c) Is there a correspondence in the appearance of the control charts for Diaper Weight, Diaper Bulk and Diaper Density? Would you expect there to be one? If so why? If not, why will they disagree?

---

6. Using the techniques of earlier exercises, investigate the effects of changing the sample size on the distribution of *MeanDensity*.

Chapter 10 "Are We In Control?"

Hypothesis Tests - One Population

Section 10.1 Overview

Statistical Objectives: After reading this chapter and doing the exercises a student will:

- Know how to set up and perform a hypothesis test involving a sample mean.
- Know when to use a Z-test and when to use a t-test.
- Know what the level of significance of a hypothesis test is and how to choose an appropriate level for a specific test.
- Know what it means to reject the null hypothesis.
- Know how to set up and perform a hypothesis test involving a sample variance.
- Know how to choose between the sample and theoretical variance.

Section 10.2 Problem Statement

In Chapter 8 we saw that the management of the tissue company was concerned about customer complaints involving sheets tearing on removal. They decided to look at the manufacturing process to see how it compared to the product specifications and to see if any changes needed to be made.

So far the engineers looking at the process have been able to get a good picture of what the actual process is doing and compare that, visually and empirically, to the specifications. Now they need to know whether they should adjust the process. Making changes to the manufacturing process is a BIG job and before they proceed they would like to be a little more certain that the changes need to be made. They decide that they will perform hypothesis tests on the data they have collected to see if the adjustments are really necessary.

Section 10.3 Characteristics of the Data Set

FILENAME: Ch10Dat.xls   An Excel Workbook
SIZE: COLUMN   3
             ROWS   200

The first seven lines of the data file are shown in Figure 10.1.

	A	B	C	D	E
Ch10Dat.xls					
1	*Day*	*MDStrength*	*CDStrength*		
2	1	1006	422		
3	1	994	448		
4	1	1032	423		
5	1	875	435		
6	1	1043	445		
7	1	962	464		

Data / ZTest / TTest / Sheet2

Figure 10.1 A Fragment of the Tissue Strength Data File

Notes on the data file:

1. The variable *Day* keeps track of the day on which the sample was taken and goes from 1 to 3.

2. The variable *MDStrength* measures Machine-directional Strength and is measured in lb./ream.

3. The variable *CDStrength* measures Cross-directional Strength and is measured in lb./ream.

Open

Click the Open *button or use the* File | Open *command (*ALT*+*F*,* O*) to open the file.*

Open the data set named **Ch10Dat.xls** from your data disk. Note that Procedure 3.1, on page 33, contains detail instruction about how to open an Excel workbook.

Section 10.4 Testing a Hypothesis

In Chapter 8 you *speculated* about whether the sample data from the tissue process indicated that the process was meeting specifications. If the company is going to make a decision about whether or not to adjust the process or to change the specifications, they will need a little more than just speculation!

In order to decide whether the sample data for *MDStrength* meets the specifications, we will have to **test a hypothesis.** The first step in testing a hypothesis is deciding what the appropriate hypothesis is. The product specifications for **MD Strength** state that the measurement is normally distributed with $\mu = \textbf{1000}$ and $\sigma = \textbf{50}$. You want to decide whether your sample came from a population with these parameters. That is, you want to decide whether there is any evidence that the sample data does not match the hypothesized product specifications. We can then state the hypothesis for the test as:

Null Hypothesis (H_O): $\mu = 1000$

Alternative Hypothesis (H_A): $\mu \neq 1000$

After setting up the hypothesis you need to choose a level of significance, α, for the test. For this example you will test at a **0.05** level of significance. Section 10.7.1 discusses the effects of the choice of significance level. The next step is to decide which type of hypothesis test to use, **Z** or **t**. Usually decisions on the type of hypothesis test about a mean are based on two things; sample size and whether you are using a sample or population variance. Table 10.1 lists the criteria for each of the tests.

Population	Variance	Sample Size	Test
Normal	Known	$n \geq 0$	Z - test
Normal	Unknown	$n \geq 30$	Z - test
Normal	Unknown	$n < 30$	t - test
Non-normal	Known	$n \geq 30$	Z - test
Non-normal	Unknown	$n < 30$	non-parametric

Table 10.1 Criteria for Z and t tests

For your test you have **200** observations in the sample. From the table you can see that when **n ≥ 30** and you are using a known (hypothesized) variance the test of choice is the **Z** - test regardless of the distribution of the population. In this case, your work in Chapter 7 indicated that the assumption of normality did not appear to be too far off base.

Section 10.5 Hypothesis Testing in Excel

Figure 10.2 shows a **Z**-test model for a single mean with respect to three typical alternative hypotheses. Simply switch to sheet **Z-Test** and enter appropriate values or formulas into the **Input** section (range **C3:C7**).

	A	B	C
	Ch10Dat.xls		
1	**Hypothesis Test for Mean - Z-test**		
2	**Input**		
3	Sample Mean	m	52.5
4	Hypothesized Mean	μ_0	50
5	Standard Deviation	σ	10
6	Sample Size	n	36
7	Significance Level	α	0.05
8	**Output**		
9	Standard Error	StdErr	1.6667
10	Z-statistic	z	1.5000
11	**Lower-Tail Test for H_A: $\mu < \mu_0$**		
12	Lower-Tail Critical Z	Z_L	-1.6449
13	Lower-Tail p-value	p-value$_L$	0.9332
14	**Two-Tail Test for H_A: $\mu \neq \mu_0$**		
15	Two-Tail Absolute Critical Z	Z_T	1.9600
16	Two-Tail p-value	p-value$_T$	0.1336
17	**Upper-Tail Test for H_A: $\mu > \mu_0$**		
18	Upper-Tail Critical Z	Z_U	1.6449
19	Upper-Tail p-value	p-value$_U$	0.0668

| | Data \ **ZTest** \ TTest \ Sheet |

Figure 10.2 Typical Z-test Cases for a Single-Mean

In order to adopt this model to testing the mean of the variable *MDStrength*, enter the following formulas:

Procedure 10.1 Providing Input for the Z-test Model

	Task Description	Mouse/Keyboard
❶	Calculate the mean of **MDStrength**: ♦ Switch to sheet **Z-test**, ♦ Select cell **C3** and ♦ Enter the =**Average(MDStrength)** function.	[image of data sheet tabs: 21 / 1 / 91.; 22 / 1 / 873; Data / ZTest] [F5] C3 [Enter] =Average(MDStrength) [↓]
❷	To define the hypothesized mean, enter **1000** into cell **C4**.	1000 [↓]
❸	To provide the known standard deviation, enter **50** into cell **C5**.	50 [↓]
❹	To specify the sample size, enter the =**Count(MDStrength)** function into cell **C6**.	=Count(MDStrength) [↓]
❶	To determine the significance level, enter **5%** into cell **C7**.	5% [Enter]

Note:
The variables on the **Data** *sheet are already named using the labels stored in the first row.*

Note:
If the third criterion demonstrated in Table 10.1 applied, you could define the standard deviation in cell **C5** *by the* =**StDev(MDStrength)** *function rather than by value.*

Figure 10.3 shows the test outcome for the **MDStrength** variable.

Section 10.6 Interpreting the Test Output

Once the mechanics of the test procedure are completed you need to look at the results and make a decision about your hypotheses. The first step is to examine the output carefully to find the values necessary to make the decision.

	A	B	C
1	**Hypothesis Test for Mean - Z-test**		
2	**Input**		
3	Sample Mean	m	989.3822
4	Hypothesized Mean	μ_0	1000
5	Standard Deviation	σ	50
6	Sample Size	n	225
7	Significance Level	α	5%
8	**Output**		
9	Standard Error	StdErr	3.3333
10	Z-statistic	z	-3.1853
11	**Lower-Tail Test for H_A: $\mu < \mu_0$**		
12	Lower-Tail Critical Z	Z_L	-1.6449
13	Lower-Tail p-value	p-value$_L$	0.0007
14	**Two-Tail Test for H_A: $\mu \neq \mu_0$**		
15	Two-Tail Absolute Critical Z	Z_T	1.9600
16	Two-Tail p-value	p-value$_T$	0.0014
17	**Upper-Tail Test for H_A: $\mu > \mu_0$**		
18	Upper-Tail Critical Z	Z_U	1.6449
19	Upper-Tail p-value	p-value$_U$	0.9993

`|◄| ◄ | ► |►|\ Data \ZTest / TTest / Sheet2 |◄|`

Figure 10.3 Z-Test Output for the Mean of
MDStrength

Reminder:
***Outside** the critical value means that the test statistic is greater than the positive critical value or less than the negative critical value.*

There are two ways to make the decision in a hypothesis test. When you do it "by hand" you look at the value of the **z** statistic and compare it to the appropriate critical value(Z_L, Z_T, or Z_U). If the **z** statistic is "outside" the critical value then you will reject H_0 and conclude that H_A is true. If the **z** statistic is "inside" the critical value then you cannot reject H_0. That is, there is no evidence that it is false.

In Figure 10.3, you can see that the **z** statistic for the sample data is **-3.1853**. In order to make the comparison, you need to know the critical values for the level of significance, α, of the test. Remember that our choice was **0.05** (or **5%**). For a two sided (**Two-Tail**) test with $\alpha = 0.05$, the lower critical value is equal **-1.96** and the upper critical value is equal **1.96**. (Notice that the model shows only the absolute value in cell **C15**). Each of the critical values is associated with tail probability of **0.025** ($\alpha/2$), $P(Z \leq -1.96) = P(Z > 1.96)$.

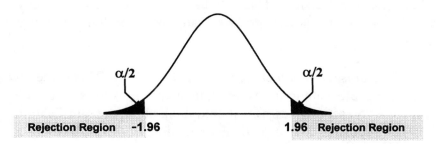

Figure 10.4 Critical Values of the Hypothesis Test

Because **-3.1853** is less than the lower critical value (**-1.96**), you reject H_0 and conclude that the sample data did not come from a normal population with a mean of **1000** and a standard deviation of **50**.

Along with the test statistic and the critical values for the test, you see something called the "**p-value**" of the test (cell **C16** for the two sided test). The **p-value** of the test is another way to make the decision about H_0. You may remember that the **p-value** is also known as the observed significance level of the test. That is, it is the smallest value of α for which the test would have resulted in rejecting H_0 with the given data set. The **p-value** for this test is **0.0014**. This value is the point at which the decision about H_0 changes from reject to accept. Compare the value of α that you used for the test to the **p-value**. If the **p-value** is less than the value of α, you will reject H_0. If the **p-value** is greater, you will fail to reject H_0.

Section 10.7 After the Test - Now What?

Section 10.7.1 Changing the Alpha Level of the Test

When you set up a hypothesis test you select a value for α, the significance level of the test. Recall that α is associated with the Type I error. It is the probability that you will reject H_0 when, in fact, it is true. Most textbooks define α but spend little time talking about how you go about choosing an appropriate value.

In choosing a value for α you must consider the **cost** of making a wrong decision. That is, what will you do, if you reject **H₀** and what will it cost you (in dollars, manpower, goodwill etc.), if you are wrong.

The most common choice for α is **0.05 (5%)**. This is the level that most people want to use when doing hypothesis tests, but really every time you set up a test you should think about the ramifications of a wrong decision in that particular situation.

For the manufacturing company in question the results of incorrectly rejecting **H₀** (saying the average **MD Strength** is not **1000** when in fact it is) might take one of two forms. One result is that the machine operators will make adjustments to raise the level of **MD Strength**, since on the basis of the data it appears to be low. The other is that they will attempt to troubleshoot the process to look for the causes of low **MD Strength** only to find they are on a wild goose chase! In either case, the effect will be machine down time and perhaps defective product.

Suppose that the company in question decides that they cannot afford machine down time. They decide that they can only afford a **1%** chance of rejecting **H₀**. Changing α will change the critical value of the test and perhaps your decision.

Exercise 1. Find the critical values for a two sided test when α = **0.01**. Does your decision about **H₀** change? If so, what is the new decision?

You could answer the question directly from the **p-value** of the test. Compare the new level of significance, **0.01** to the **p-value** of **0.0014**. Since the **p-value** of **0.0014** is less than **0.01**, you will reject **H₀**.

Exercise 2. Suppose the company decides that the cost of machine down time is not great and that they can afford a larger Type I error. Select an appropriate value for α and find the critical values. What effect does this have on the decision from the original (α = **0.05**) test?

You might wonder why someone would ever want to make the possibility of an error, in this case a Type I error, **larger.** The reason is that along with an α error which you set for each test, there is another error that is associated with the test, the Type II error, β. The β error of the test is the probability that the test will fail to reject $\mathbf{H_0}$ when it is in fact false. That is, it is the probability that the experimenter will decide that there is not enough evidence to take action and maintain status quo. In the case of the tissue manufacturer, that would mean **NOT** making changes to a process that is definitely in error.

Although calculating the β error is a bit complicated, the fact is that its value is in direct tradeoff with α for a sample of a given size. That is, as the probability of making a Type I error decreases, the associated Type II error probability increases. Thus the person performing the test might decide to increase α so that the β error decreases.

Section 10.7.2 One Sided vs. Two Sided Tests

For the manufacturing company, the question of interest was whether or not the process was running to target specifications. That is, whether the mean **MD Strength** was equal to **1000** or whether it was not equal to **1000**. Thus, the test you did was a two sided test.

Suppose, however, that the company did not care if the **MD Strength** was high, but was concerned only about low **MD Strength**. That is, they would only take action, if the data showed that the values were too low to have come from the target population. In this case you would want to test the hypotheses

H_0: $\mu \geq 1000$

H_A: $\mu < 1000$

As you can see, the test setup on sheet **Z-Test** also handles this case.

Exercise 3. Perform the one sided test described above to determine whether the mean **MD Strength** is less than **1000**. What is the result?

Section 10.8 Other Considerations

At this point it is important to figure out what all this might MEAN to the manufacturing company in question. Rejecting the hypothesis means that the sample was taken when the process was not running according to the manufacturing specifications. If the sample was representative of the process at any time, then the logical conclusion is that the product they are making is not what they intend.

The natural conclusion of the test is that the mean of the process is not **1000**, but is, in fact, some other value. Thus the company should consider adjusting the process and retest before making any decisions about changing the specifications. **HOWEVER,** before you make a suggestion like that to the management, consider taking a second look at the test you performed.

Remember that you performed a hypothesis test using population (that is, assumed) values for both the mean **AND** the variance. Is it possible that the population variance is not the value that you assumed. An undetected change in the population variance can make it appear that the population mean is no longer correct.

	A	B	C
1	Hypothesis Test for Mean - Z-test		
2	Input		
3	Sample Mean	m	989.3822
4	Hypothesized Mean	μ_0	1000
5	Standard Deviation	σ	55.1814
6	Sample Size	n	225
7	Significance Level	α	5%
8	Output		
9	Standard Error	StdErr	3.6788
10	Z-statistic	z	-2.8862
11	Lower-Tail Test for $H_A: \mu < \mu_0$		
12	Lower-Tail Critical Z	Z_L	-1.6449
13	Lower-Tail p-value	p-value$_L$	0.0019
14	Two-Tail Test for $H_A: \mu \neq \mu_0$		
15	Two-Tail Absolute Critical Z	Z_T	1.9600
16	Two-Tail p-value	p-value$_T$	0.0039
17	Upper-Tail Test for $H_A: \mu > \mu_0$		
18	Upper-Tail Critical Z	Z_U	1.6449
19	Upper-Tail p-value	p-value$_U$	0.9981

Note:
*Here, the cell **C5** contains the* =StDev(MDStrength) *function.*

Figure 10.5 Z-test with the Sample Standard Deviation

On the **Z-Test** sheet, replace the value of the standard deviation with the sample standard deviation of the variable *MDStrength*. Go the cell **C5** and enter **=StDev(MDStrength)**. The outcome is shown in Figure 10.5.

Look at the value of the sample standard deviation and compare it to the assumed value of **50**. It would appear that the sample value of **55.1814** might be higher than the assumed value. What would happen if you did the test using the sample variance instead?

Exercise 4. Interpret the results of the original hypothesis test for σ equal to the sample standard deviation. Does your conclusion change?

Because you have such a large sample size the **z**-test was still the appropriate test. Although this particular change did not change the conclusion of the test, you can see that it is important that you be sure that you are setting up your tests using the correct values.

Note:
Explore the **t-Test** *sheet. It contains a solution for the test of the single mean for small samples. Check Table 10.1 to see when you would need the* **t-** *Test.*

Section 10.9 Investigative Exercises

In the following exercises you are asked to use the skills introduced in the previous chapters to extract information from the tissue strength data file. You are provided with space to answer the questions and paste in graphical output from the program. If you do not have access to a printer, you can sketch the graphs in the spaces provided.

1. After giving the matter some thought the engineers looking at the tissue manufacturing process wondered if the fact that the measurements were taken on three different days might matter. Since operating conditions are not always the same on any given day, they wondered if the process was off target on all three days or if it was only off on one or two days. They decided to check each day individually.

Perform a hypothesis test at the **0.05** level of significance to determine whether the mean **MD Strength** was **1000** on each of the three days separately. Use the target standard deviation of **50** and write your results in the table below:

DAY	Z Statistic	Decision (Reject H_0 / Fail to Reject H_0
1		
2		
3		

2. For each of the three days, find the sample standard deviation and redo the test using this value. Does the decision change? Fill in the table below.

DAY	Z Statistic	Does the Decision Change?
1		
2		
3		

3. Using the results of all of hypothesis tests that you have done, do you think that the company can assume that **MD Strength** is running to target? Why or why not?

4.a) Using a **0.05** level of significance, does the sample data indicate that the variable *CDStrength* was running according to target specifications for the three day period as a whole?
Note: **CD Strength** is supposed to be normally distributed with a mean of **450** and a standard deviation of **25**.

b) Test to see if the use of the sample standard deviation changes the decisions.

---

5. Since the consumer problem that prompted the study was sheets tearing on removal, the company decides that it is really only interested in knowing whether the average **CD Strength** is less than the target of **450**. If it is not, they will not make any adjustments. Redo the test of the previous exercise as a one-sided test to reflect this. What is the result?

---

6. Judging the results of the tests you have done with **CD Strength**, do you see any need for the group to test each of the three days separately? Why or why not? What would you expect to happen if you did test the days separately?

7. The group is divided on the subject of **GMT**. The operations specialists think that since the company has critical specifications for the measurement they need to test to see if it is running according to those specifications. The machine operators say that since the **GMT** is calculated from the other two variables, which have been tested, they do not need to perform any hypothesis tests concerning this data. Which group do you agree with and why?

8. Using the results of your analysis prepare a report to management telling them about the current tissue manufacturing process. Make recommendations to them on whether the process needs to be adjusted or whether it is running to target. Remember, if it is running to target they will be considering making changes to the specifications to reduce customer complaints about dispensing. Address this point in your report. Compare your conclusions now to the more subjective ones you put forth in Chapter 8.

Chapter 11 "Will the Wrapper Fit?"

Hypothesis Tests - Two Populations

Section 11.1 Overview

Statistical Objectives: After reading this chapter and doing the exercises a student will:

- Know how to set up the hypotheses for a two-population test concerning sample means.
- Know how to decide which two-sample test is appropriate for a given situation.
- Know when the assumption of equal variances is appropriate.

Section 11.2 Problem Statement

A manufacturing company that makes paper products is having trouble with their paper towel line. When a roll of paper towels is manufactured, the last step before the roll is put in a case is to wrap it with the poly wrapper that the consumer sees it in on the store shelf. A large amount of product is being scrapped because it is not being wrapped properly.

The problem appears to be that the diameters of the towel rolls are too large and the wrapper does not go all the way around and seal properly. There are many factors in the manufacturing process that would appear to affect roll diameter, however the engineers involved were not sure exactly how different machine settings really impacted roll diameter. In fact, they were not convinced that all of the factors made a difference!

The towel machine team designed a study to determine the extent that different machine settings actually had on roll diameter. Several people on the team expressed a concern that some settings, which would result in a good roll diameter, might have an adverse impact on another towel roll characteristic, roll firmness.

If a towel roll is not firm that will also affect wrapping, and if it is too firm consumers will perceive it as stiff and react negatively. Fixing the roll diameter problem at the expense of firmness was not an option.

The team decided to look at three machine factors:

1. **Embosser roll gap**: the mechanism that puts the pattern in the towel
2. **Draw roll gap**: the opening through which the towel material is pulled onto the winders
3. **Speed**: the speed at which the machine winds the rolls of towels.

The machine was run at different settings for each factor and rolls of towels were taken from the end of the production line. These towel rolls were measured on two characteristics, roll diameter and roll firmness.

Section 11.3 Characteristics of the Data Set

FILENAME: Ch11Dat.xls   An Excel Workbook
SIZE: COLUMN   5
             ROWS   75

The first seven lines of the data file are shown in Figure 11.1.

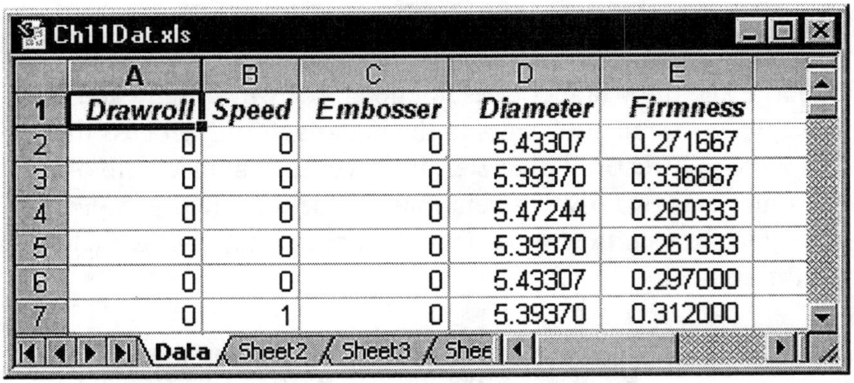

	Drawroll	Speed	Embosser	Diameter	Firmness
1	Drawroll	Speed	Embosser	Diameter	Firmness
2	0	0	0	5.43307	0.271667
3	0	0	0	5.39370	0.336667
4	0	0	0	5.47244	0.260333
5	0	0	0	5.39370	0.261333
6	0	0	0	5.43307	0.297000
7	0	1	0	5.39370	0.312000

Ch11Dat.xls

Data / Sheet2 / Sheet3 / Shee

Figure 11.1 The Towel Diameter Data File

Notes on the Data File:

1. The variable *Drawroll* is a **0-1** variable that indicates the size of the drawroll gap. A **0** indicates that the smaller gap measurement was used, while a **1** indicates the larger gap measurement.

2. The variable *Speed* is a **0-1** variable that indicates the machine speed. A **0** indicates the slower machine speed was used and a **1** indicates that the faster machine speed.

3. The variable *Embosser* is a **0-1** variable that indicates the size of the embosser roll gap. A **0** indicates that the smaller gap measurement was used, while a **1** indicates that the larger gap measurement was used.

4. The variable *Diameter* measures the diameter of the roll of paper towels in inches.

5. The variable *Firmness* measures the firmness of the roll of paper towels on a specialized scale. Lower numbers indicate softer rolls while larger numbers indicate firmer rolls.

Open the data set named **Ch11Dat.xls** from your data disk. Note that Procedure 3.1, on page 33, contains detail instruction about how to open an Excel workbook.

Click the Open *button or use the* File | Open *command (*ALT+F, O*) to open the file.*

Section 11.4 Hypothesis Testing - Two Populations

Section 11.4.1 Setting Up the Hypotheses

The questions that the towel team wants answered are about whether certain changes to the machine settings result in changes to two towel characteristics, mean roll diameter and mean firmness. They looked at the machines under two conditions for each machine setting and sampled product from each set of conditions. Thus they are interested in testing whether the mean towel characteristic is the same for each **population** (machine setting).   You can state the hypothesis for this test as:

$$H_O: \quad \mu_1 = \mu_2 \qquad \text{OR} \qquad \mu_1 - \mu_2 = 0$$
$$H_A: \quad \mu_1 \neq \mu_2 \qquad \qquad \qquad \mu_1 - \mu_2 \neq 0$$

The first item that they decided to look at was whether there was a difference in roll diameter when the machine was run at two different speeds.

Exercise 1. Create a separate column for the variable *Diameter* for each value of the variable *Speed*.

Note:
Procedure 5.9 (page 118) shows detail steps about extracting data from a data list. Procedure 5.10 (page 121) shows how to copy data from one sheet to another.

Hint: On the **Data** sheet, create the **Criteria** range (**G1:G2**) for *Speed* = **0** and the **Output** range (**G4**). Select any cell within the data range, invoke the **Data | Filter | Advanced Filter** command and set it all up as shown in Figure 11.2. Copy the **Output** range (**G4:G29**) to **sheet2**, starting at **A1**. Next, get back to the **Data** sheet, set the *Speed* value in **G2** to **1**, extract the other *Diameter* data, and again copy the **Output** range (**G4:G29**) to **sheet2**, this time starting at **B1**.

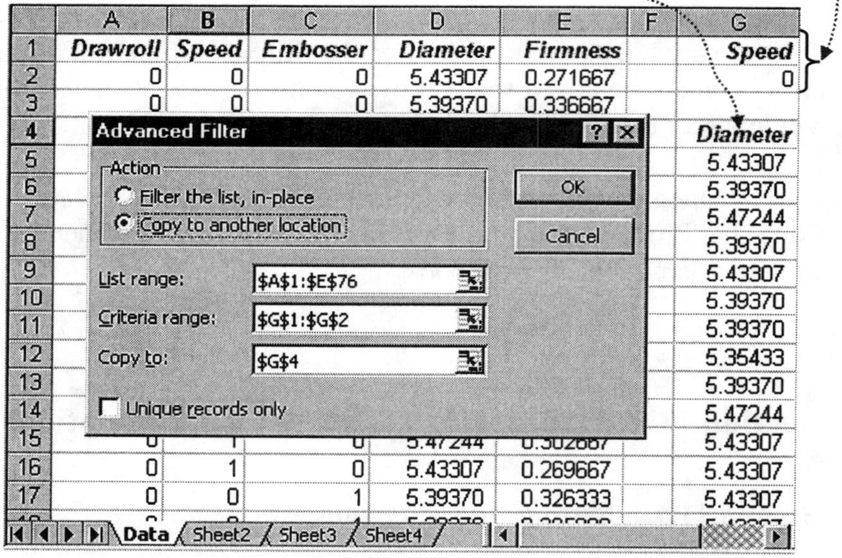

Figure 11.2 Extracting the Subsets of the Variable Diameter for Speed=0

Finally, rename the labels in **A1** and **B1** on sheet **2** as **Diameter0** and **Diameter1**, and change the sheet tab name to **DiameterBySpeed**.

Figure 11.3 shows the two *Diameter* subsets, one for *Speed*=**0** and the other for *Speed*=**1**.

Section 11.4.2 Choosing the Correct Test

Before you rush headlong into any statistical tests, it would be wise to **look** at the data you are about to analyze so that you have a sense of what the test results are all about. Usually you should obtain a set of descriptive statistics and a graphical display of the data. This is also useful when you need to consider the assumptions in certain statistical tests. Looking at the variable *Speed* you see that it has two values **0** and **1**. Each of these values identifies the population from which the associated sample was taken.

	A	B	C
1	*Diameter0*	*Diameter1*	
2	5.43307	5.39370	
3	5.39370	5.47244	
4	5.47244	5.43307	
5	5.39370	5.47244	
6	5.43307	5.43307	
7	5.39370	5.51181	
8	5.39370	5.47244	
	....??	5.47244	
49		5.47244	
50		5.43307	
51		5.39370	

Figure 11.3 Subsets of Diameter for Speed=0 and Speed = 1

Exercise 2. Create a set of summary statistics for the variables *Diameter0* and *Diameter1*.

The basic summary statistics for each sample are shown in Figure 11.4.

	A	B	C	D	E	F
1	*Diameter0*	*Diameter1*			*Diameter0*	*Diameter1*
2	5.43307	5.39370		Mean	5.3952748	5.424409
3	5.39370	5.47244		Standard Error	0.00834545	0.008734
4	5.47244	5.43307		Median	5.3937	5.43307
5	5.39370	5.47244		Mode	5.35433	5.43307
6	5.43307	5.43307		Standard Deviation	0.041727248	0.06176
7	5.39370	5.51181		Sample Variance	0.001741163	0.003814
8	5.39370	5.47244		Kurtosis	-0.77930371	2.240966
9	5.35433	5.47244		Skewness	0.143151289	-1.53581
10	5.39370	5.47244		Range	0.15748	0.27559
11	5.47244	5.43307		Minimum	5.31496	5.23622
12	5.43307	5.47244		Maximum	5.47244	5.51181
13	5.43307	5.47244		Sum	134.88187	271.2204
14	5.43307	5.47244		Count	25	50

Figure 11.4 Summary Statistics for Diameter by Machine Speed

Note:
Chapter 6 explores the summary measures in depth. In particular, Procedure 6.2 (page 137) shows how to generate basic summary measures.

Note:
This report was produced by the Tools | Data Analysis | Descriptive Statistics command and the following specifications:
  Input Range: $A$2:$B$51
  Output Range: $D$1
  ☑ **Summary Statistics**
The output was slightly edited.

Exercise 3. Create a histogram of the variables *Diameter0* and *Diameter1*.

Hint:
*To generate the frequency table, first type the bin values as shown in Figure 11.5. For **Diameter0**, select range I4:I8, type =Frequency(A2:A26, H4:H8), and press* CTRL + SHIFT + ENTER.

*For **Diameter1**, select range J4:J8, type =Frequency(B2:B51, H4:H8), and press* CTRL + SHIFT + ENTER.

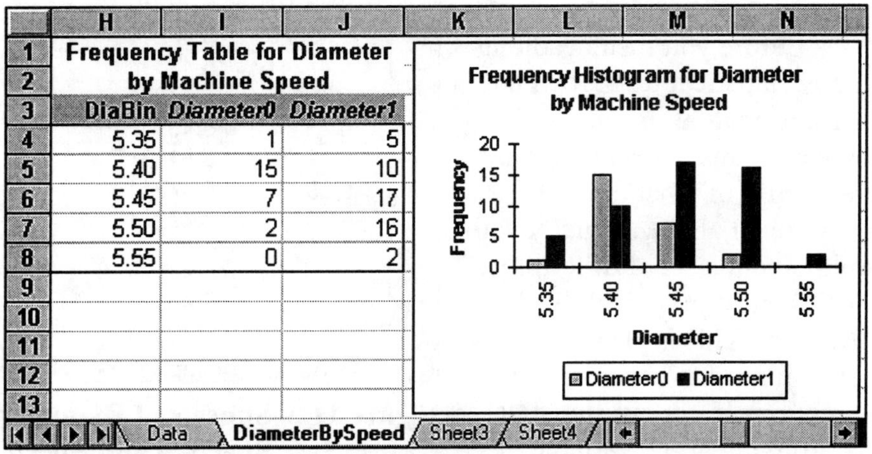

Figure 11.5 Frequency Distribution of Diameter by Machine Speed

Note:
Procedure 5.13 (page 125) shows how to generate frequency histograms for two data sets.

The frequency distribution (table and histogram) for each sample are shown in Figure 11.5. Notice that the frequency table was created using the **Frequency()** array-function. The histogram was generated based on the frequency table using the **ChartWizard**.

From the summary statistics you can see that the average roll diameter is larger for the higher machine speed, but is it a *significant* difference (that is, one that did not happen by chance)? To determine this you will have to perform a hypothesis test.

The first step in doing the test is to determine which test, **Z** or **t**, you should do. Table 11.1 lists the criteria for each of the tests.

Populations	Variances	Sample Sizes	Test
Independent Normal	Known	$n_1 \geq 0, n_2 \geq 0$	Z-test
Independent Non-Normal	Unknown	$n_1 \geq 30 , n_2 \geq 30$	Z- test
Independent Normal	Unknown but assumed equal	$n_1 \leq 30, n_2 \leq 30$	t-test (Pooled variance)
Independent Normal	Unknown and assumed not equal	$n_1 \leq 30, n_2 \leq 30$	Modified t-test (Behrens - Fisher problem)

Table 11.1 Table of Test Criteria

Looking at the summary statistics you can see that the sample sizes are not both $\geq$ 30, and so you will need to use a **t**-test. Looking at Table 11.1, you also see that using a **t**-test requires that the populations from which the samples were taken be *independent and normally distributed* and that choice of a test depends on whether the sample variances can be assumed to be equal.

To determine whether or not the data are normally distributed you can look at the histograms of the data that you just created. While this is only an approximate test, unless there are large departures from normality (very skewed or non-mound shaped distributions) the test will be valid. You could also apply the empirical rule for normality (see Chapter 8, if you do not remember how to do this). The distributions of the data shown in Figure 11.5 are mound shaped and while they are not ideally symmetric, it would appear that the assumption of normality is not violated.

If you look at the data in Figure 11.2 you see that the sample standard deviation for *Diameter* with *Speed* = 0 is **0.0417** in. while the sample standard deviation for *Speed* = 1 is **0.0618** in. (The difference in variability is also evident in the histograms.) They look different but to really need to know whether they are statistically different you would have to do another statistical test. We will address this issue in the next section.

Section 11.4.3 Doing the Hypothesis Test

The following procedure will guide you through the mysterious steps of doing the **t-Test** to find out whether or not the *Diameter* means for different *Speed* (**0** and **1**) are statistically different.

Procedure 11.1 Doing a t-Test for Two Samples with Unequal Variances

Note:
Use the **Tools | Data Analysis | F-Test Two-Sample for Variances** *command to test if the variances are or are not equal.*

	Task Description	Mouse/Keyboard
❶	Invoke the ◆ **Tools** ◆ **Data Analysis** command	Tools ABC Spelli... F7 Wizard Data Analysis...

Excel opens the **Data Analysis** *dialog box.*

*Now you are faced with an important decision. Remember that our choice of test depended on whether we assumed that the sample variances are equal to each other. There is a test, called an F test, to actually determine this, but it is not covered in most basic statistics courses. As a rule, unless you **KNOW** your variances are equal, it is safer to assume that they are not. The test is sensitive to this assumption and the test results that you get could be in error if they are not equal and you assume that they are. Since our standard deviations are not the same (**0.0417** in. vs. **0.0618** in.) we will not assume the variance are equal.*

---

❷

◆ Scroll down the **Analysis Tools** list to reveal the
t-Test: Two Sample Assuming Unequal Variances option.

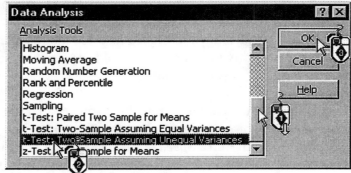

◆ Select this option, and
◆ Click the **OK** button.

Excel opens the **t-Test: Two Sample Assuming Unequal Variances** *dialog box.*

---

❸

◆ In the **Variable 1 Range** box, specify the range of the *Diameter0* variable. Type or highlight range **A2:A26**.

◆ In the **Variable 2 Range** box, specify the range of the *Diameter1* variable. Type or highlight range **B2:B51**.

◆ For the **Hypothesized Mean Difference**, type **0** (to find out whether or not the means are equal).

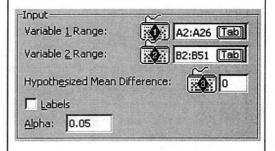

④	◆ Click the **Output Range** option button and, ◆ in the adjacent box, type or point to cell **D17**, and press ⌈ENTER⌉.	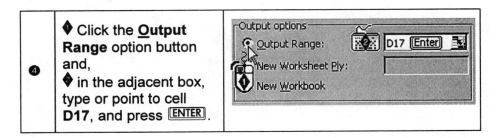

Excel tests the null hypothesis that the mean of the first sample is equal to the mean of the second sample at the significance level $\alpha = 0.05$ (confidence level, $1-\alpha = 0.95$). Excel generates the test's output starting at the cell that you have identified in the **Output Range** box. Figure 11.6 shows the test outcome.

	D	E	F
17	t-Test: Two-Sample Assuming Unequal Variances		
18			
19		Variable 1	Variable 2
20	Mean	5.3952748	5.4244086
21	Variance	0.001741163	0.003814258
22	Observations	25	50
23	Hypothesized Mean Difference	0	
24	df	66	
25	t Stat	-2.41169477	
26	P(T<=t) one-tail	0.009334018	
27	t Critical one-tail	1.668270215	
28	P(T<=t) two-tail	0.018668035	
29	t Critical two-tail	1.996563697	

Figure 11.6   Results of the t-Test for Two Means

Section 11.4.4 Analyzing the Output

Excel provides the 2-Sample **t** test outcomes for both the one-sided (one-tail) and two-sided (two-tail) cases. The degrees of freedom (**df**) and the test statistic (**t Stat**) are calculated based on the following formulas:

$$t = \left(\overline{x}_A + \overline{x}_B - \Delta \right) / \sqrt{\frac{s_A^2}{n_A} + \frac{s_B^2}{n_B}}$$

$$df = \frac{\left(\dfrac{s_A^2}{n_A} + \dfrac{s_B^2}{n_B}\right)^2}{\dfrac{\left(s_A^2/n_A\right)^2}{n_A - 1} + \dfrac{\left(s_B^2/n_B\right)^2}{n_B - 1}}$$

where:

$\bar{x}_A, s_A^2, n_A$ are the mean, variance and size of sample A,

$\bar{x}_B, s_B^2, n_B$ are the mean, variance and size of sample B,

Δ is the hypothesized difference between the means.

Note that all these values are provided in the first four rows of the test report (see Figure 11.6).

The value of the **t**-statistic is **-2.41**. For the two-sided test, Excel provides only the absolute critical value (here **1.9966**). Comparing the test statistic to both the critical values **-1.9966** and **1.9966** you find that it falls into the rejection region (below the lower limit). Thus you reject H_O and conclude that mean *Diameter* is significantly different for the two speeds.

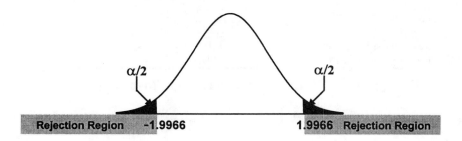

Figure 11.7   Critical Values for the Two-tail t Test

You could also reach this decision by looking at the **p-value** for the test and comparing it to the chosen level of significance. In this case since the **p-value = 0.01867** is less than the value of α you reject H_O.

Exercise 4. Suppose you changed the value of α for the test from **0.05** to **0.01**. How do the conclusions change?

Section 11.4.5 Statistical Differences vs. Meaningful Differences

Although the difference in the mean values of *Diameter* is *statistically* significant you do not know if the difference is meaningful to the manufacturing company. Very often statistical significance and practical significance are not the same thing. The value of the difference in the two means is **0.0291** inches.

By now you probably have noticed that the values of *Diameter* are a bit strange. Although it appears that *Diameter* is measured to **4** decimal places, there are not as many *different* values for *Diameter* as you would expect. In fact, there are only **8** different values that the **4** decimal places can be. Can you think of a reason that this might happen?

The real reason is that the value of *Diameter* was measured to the nearest millimeter and converted to inches because the product specifications for the paper towel product are measured in English measures, not metric. The converted values were recorded with four decimal places. Since the original values were rounded to the nearest millimeter, the converted values must differ by exactly **0.0394** (using the conversion factor of **25.4** millimeters per inch).

Thus the statistically significant difference of **0.0291** inches is *smaller* than the smallest difference that the measuring devices could detect! This is an example of how statistically different and practically different are **NOT** necessarily the same. It is also an example of how the accuracy of the measuring device can affect results.

Section 11.5 Investigative Exercises

In the following exercises you are asked to use the skills introduced in the previous chapters to extract information from the tissue strength data file. You are provided with space to answer the questions and paste in graphical output from the program. If you do not have access to a printer, you can sketch the graphs in the spaces provided.

1. Perform a hypothesis test at the **0.05** level of significance to determine whether the mean value of *Firmness* differs when the speeds are different.

a) Were you willing to assume that the two population variances were equal?

b) Was there a significant difference in the mean *Firmness* for the two populations?

c) Change the value of alpha to **0.01**. How does this change the results of the test? What if you change it to **0.10**?

---

2. a) Using the procedure outlined in the chapter, set up and perform hypothesis tests to determine whether there is a difference in mean *Diameter* for the two populations defined by *Drawroll*. Do the same thing for *Embosser*. Use a level of significance of **0.01**. Fill in the table below:

Population Factor	Sample Sizes	Assume Variances Equal?	Decision Reject H_0/ Fail to Reject H_0
Drawroll			
Embosser			

b) If the differences in the mean *Diameter* are statistically significant, are they of practical significance?

3.a) Repeat the previous exercise for the variable *Firmness*.

Population Factor	Sample Sizes	Assume Variances Equal?	Decision Reject H_0/ Fail to Reject H_0
Drawroll			
Embosser			

b) Based on these results would you say that there is a significant difference in mean *Firmness* for the two populations defined by *Drawroll*?

c) What about the populations defined by *Embosser*?

4. The towel team is also interested in knowing whether *Diameter* meets the specifications for the roll of towels, **5.35** inches.

a) Perform a hypothesis test to decide whether average *Diameter* for *Speed* = **0** meets the specifications. What test did you use and what is your conclusion?

b) Do the same thing for *Speed* = **1**. What test did you use and what is your conclusion?

c) In reality the team is interested in knowing whether the specification is met over the entire range of speeds that are used. Perform a hypothesis test to determine whether *Diameter* meets the specifications for the two populations together. What is your conclusion?

5. Repeat the previous exercise for the variable *Drawroll*.

a)

b)

c)

---

6. Repeat the exercise for *Embosser*.

a)

b)

c)

---

7. Prepare a report for management explaining the effects that each of the machine settings have on *Diameter* and *Firmness*. Indicate which settings result in acceptable values for roll diameter. Make a recommendation on machine settings if you can.

Chapter 12 "Are Your Members Happy?"

Testing Proportions

Section 12.1 Overview

Statistical Objectives: After reading this chapter and doing the exercises a student will:

- Know how to analyze survey data in order to estimate true population proportions.
- Know how to decide whether to use a one-tail test or a two-tail test.
- Understand **p-values**, how they are used and how they relate to alpha.
- Know what it means to test the **p-value** against **0.50**.
- Know what the phrase "significantly different" means.

Section 12.2 Problem Statement

Many Americans have taken up golfing. This population of golfers influences several different industries. As we saw in Chapter 5 many golf manufacturers are interested in satisfying their customers by designing golf balls which will fly further. In addition to having an influence on manufacturers, these golfers are important to another sector of the market: country clubs. Many golfers belong to a country club in order to be sure that they can get on the course without a long wait.

In addition to spending time on the golf course, many members also eat meals and socialize at their country club. Thus, if you are on the Board of Directors of such a club, you should be concerned with keeping your members satisfied. One such New England Country Club decided to survey its **250** members. They were interested in member satisfaction with the overall club conditions, the food services and the dues structure. A questionnaire was designed and mailed to all of the club members. Of the **250** current members, **134** returned the survey. This translates to a **53.6%** response rate, which is a stronger response than you usually receive from a mail survey.

The data set you will be analyzing in this chapter contains demographic information about the respondent as well as his/her answers to the questions about member satisfaction.

Section 12.3 Characteristics of the Data Set

FILENAME:	Ch12Dat.xls	An Excel Workbook
SIZE	COLUMNS	15
	ROWS	134

The first 8 columns and 7 rows of the actual data file are shown in Figure 12.1.

	A	B	C	D	E	F	G	H
	Howlong	**Type**	**Sex**	**MStatus**	**Depends**	**Age**	**Income**	**Area**
2	1	1	2	1	0	2	2	2
3	2	2	2	2	2	3	4	1
4	1	1	1	1	0	2	1	2
5	4	2	2	2	0	4	0	2
6	1	2	1	2	0	4	0	2
7	3	2	2	2	0	3	2	2

**Figure 12.1   A Fragment of the Country Club Survey
Results (columns 1-8)**

The remaining 7 columns are shown in Figure 12.2.

	I	J	K	L	M	N	O	P
	Often	**Friend**	**Condition**	**Greens**	**Landscap**	**Parking**	**Locker**	
2	3	3	2	2	3	1	2	
3	2	4	1	2	2	2	3	
4	1	2	3	3	3	2	2	
5	3	1	3	3	4	3	3	
6	2	1	3	2	3	3	2	
7	3	2	3	3	1	3	3	

**Figure 12.2   A Fragment of the Country Club Survey
Results (columns 9-15)**

Notes on the data set: Non-response is coded zero (0) for all variables.

1. The variable *Howlong* indicates how long the person has been a member of the Country Club:
>   1= under 1 year, 2= 1-3 years, 3= 4-7 years,
>   4= 8-10 years, 5= more than 10 years.

2. The variable *Type* indicates the type of membership that the member currently has:
>   1= Full, 2=Family, 3= Associate, 4= Junior, 5=College,
>   6=Social, 7= Restricted Family, 8= Senior Full, 9= Senior
>   Associate, 10= Senior Family, 11= Restricted Senior Family.

3. The variable *Sex* indicates the sex of the member:
>   1= Male, 2= Female

4. The variable *MStatus* indicates the marital status of the respondent:
>   1= Single, 2= Married

5. The variable *Depends* indicates how many dependents the member has.

6. The variable *Age* contains information about the respondents age:
>   1= under 20   2= 21-40   3=41-60   4= over 60.

7. The variable *Income* indicates the annual income range of the member:
>   1= under $25,000   2= $25,001 - 50,000
>   3= $50,001-75,000   4= $75,001 or more.

8. The variable *Area* tells if the member has ever belonged to any other country club in the area: 1= Yes, 2 = No.

9. The variable *Often* indicates how often the respondent uses the golf course:
>   1= less than once a week,   2= 1-2 times a week,
>   3= 3-4 times a week, 4= more than 4 times a week.

10. The variable *Friend* contains information regarding how often the member brings a guest to play golf at the club:

1= Always, 2= Frequently, 3= Sometimes, 4= Never.
The last 5 variables indicate the respondents ranking of 5 features of the club. The following numeric rating scheme is used:
1=Poor, 2= Fair, 3= Good, 4= Excellent.

11. The variable *Condition* is the respondents rating of the condition of the golf course.

12. The variable *Greens* is the respondents rating of the condition of the greens.

13. The variable *Landscap* is the respondents rating of the landscape surrounding the golf course.

14. The variable *Parking* is the respondents rating of the accessibility of parking.

15. The variable *Locker* is the respondents rating of the conditions of the locker rooms.

Click the **Open** *button or use the* F*ile | Open command (* ALT *+* F *,* O *) to open the file.*

Open the data set named **Ch12Dat.xls** from your data disk. Note that Procedure 3.1, on page 33, contains detail instruction about how to open an Excel workbook.

Section 12.4 Creating Cross Tabulation Tables in Excel

In order to perform a hypothesis test on proportions using Excel, you must provide the program with sample proportions. For example, the sample proportion of interest might be the proportion of respondents who ranked the condition of the greens as "excellent". This would be calculated by finding the number of respondents who ranked the condition of the greens as "excellent" and dividing by the total number of respondents.

If your data set contains raw survey data like the one in this chapter, then you must first tabulate the data in order to obtain the sample proportion. Thus, before you learn how to run hypothesis tests on proportions you must learn how to create the appropriate tables. In

Chapter 7, you learned how to use the **Pivot Table Wizard** to generate a cross-tab table to find out counts of all categories of the variable *Problem* for each category of the variable *Size*. Based on those counts, you were able to determine all relevant proportions. Here you will perform a similar task. This time, however, you will use the **Pivot Table Wizard** to generate proportions for all combinations of the categories of *Condition* and *Sex*.

Procedure 12.1 Generating Cross-Tabulated Proportions

	Task Description	Mouse/Keyboard
❶	On the **Data** sheet, select any cell within the data range.	
❷	Invoke the ♦ **Data** ♦ **Pivot Table Report** command.	
❸	With the **Microsoft Excel list or database** option selected, click the **Next** > command button.	
❹	Based on the current cell, Excel defines the range including the entire data set. Thus click the **Next** > command button.	

Excel opens the **Pivot Table Wizard - Step 3 of 4** *dialog box. Notice that all variable names of the data set are included as buttons on the right side of the dialog box.*

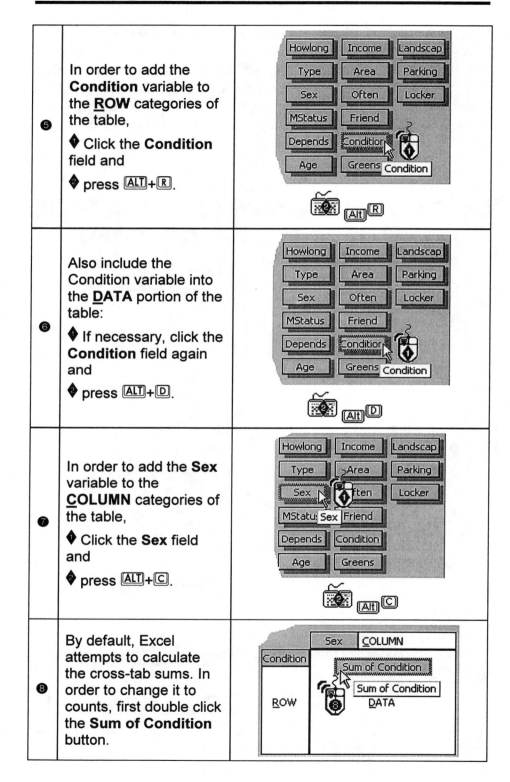

<table>
<tr>
<td>

Note:
An alternative action
would be to drag the
Condition *field onto*
ROW.

</td>
<td>

❺

</td>
<td>

In order to add the
Condition variable to
the **ROW** categories of
the table,

◆ Click the **Condition**
field and

◆ press ⟨ALT⟩+⟨R⟩.

</td>
</tr>
<tr>
<td>

Note:
An alternative action
would be to drag the
Condition *field onto*
DATA.

</td>
<td>

❻

</td>
<td>

Also include the
Condition variable into
the **DATA** portion of the
table:

◆ If necessary, click the
Condition field again
and

◆ press ⟨ALT⟩+⟨D⟩.

</td>
</tr>
<tr>
<td>

Note:
An alternative action
would be to drag the
Sex *field onto*
COLUMN.

</td>
<td>

❼

</td>
<td>

In order to add the **Sex**
variable to the
COLUMN categories of
the table,

◆ Click the **Sex** field
and

◆ press ⟨ALT⟩+⟨C⟩.

</td>
</tr>
<tr>
<td></td>
<td>

❽

</td>
<td>

By default, Excel
attempts to calculate
the cross-tab sums. In
order to change it to
counts, first double click
the **Sum of Condition**
button.

</td>
</tr>
</table>

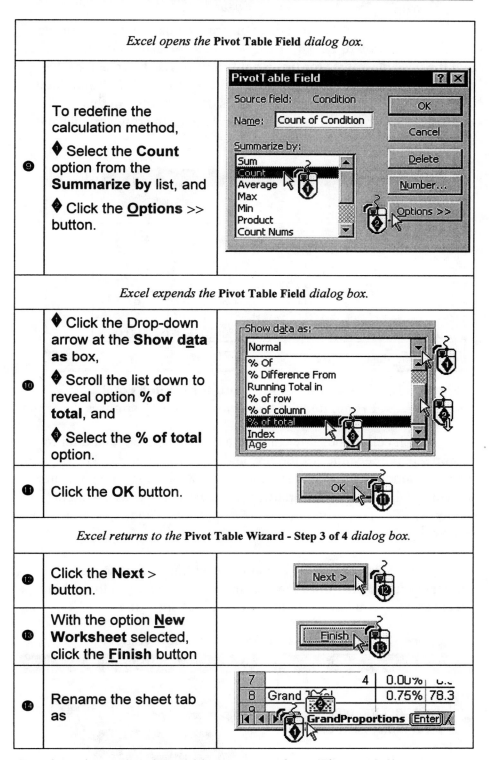

Excel opens the **Pivot Table Field** *dialog box.*

⑨ | To redefine the calculation method,

◆ Select the **Count** option from the **Summarize by** list, and

❷ Click the **Options** >> button.

Excel expends the **Pivot Table Field** *dialog box.*

⑩ | ◆ Click the Drop-down arrow at the **Show data as** box,

◆ Scroll the list down to reveal option **% of total**, and

◆ Select the **% of total** option.

⑪ | Click the **OK** button.

Excel returns to the **Pivot Table Wizard - Step 3 of 4** *dialog box.*

⑫ | Click the **Next** > button.

⑬ | With the option **New Worksheet** selected, click the **Finish** button

⑭ | Rename the sheet tab as

Excel produces the pivot table on a new sheet (Figure 12.3).

It is a good time to save your workbook.

As you examine the table, you see the possible values for the variable *Condition* (**0,1,2,3,4**) listed down the left-hand side of the table and the possible values for the variable *Sex* (**0,1,2**) listed across the top of the table. There are also row and column labeled **Grand Total**. They provide the row and column totals, respectively. In this example, the variable *Condition* is referred to as the row variable and *Sex* is referred to as the column variable.

Note:
Zeroes represent missing data.
*For example, **Sex** = **0** means that there is no answer to the **Sex** question. From the table, we learn that **0.75%** respondents did not specify their **Sex**, and **1.49%** did not evaluate Condition.*

	A	B	C	D	E
1	Count of Condition	Sex			
2	Condition	0	1	2	Grand Total
3	0	0.00%	0.00%	1.49%	1.49%
4	1	0.00%	0.00%	0.75%	0.75%
5	2	0.75%	19.40%	5.97%	26.12%
6	3	0.00%	50.00%	11.19%	61.19%
7	4	0.00%	8.96%	1.49%	10.45%
8	Grand Total	0.75%	78.36%	20.90%	100.00%

Sheet1 / Data / Sheet2 / Shee

Figure 12.3 Cross-tabulation Result for Grand Total Proportions of *Condition* by *Sex* Responses

Each cell in the table contains a percent of the grand total for its corresponding values of the row and column variables. For example, among all respondents there were **8.96%** male respondents (coded **1**) who ranked the condition of the golf course as excellent (coded **4**). The proportion of female respondents (coded **2**) in the same *Condition* category was **1.49%**.

In order to determine proportions within particular groups corresponding to either the row variable (*Condition*) or to the column (*Sex*) variable, you have to redefine the **Count** operation.

Procedure 12.2 Modifying Existing Pivot Table Report

Note:
*If the table does not get selected, click **A1** again..*

	Task Description	Mouse/Keyboard
❶	To select the entire pivot table, click cell **A1**.	

	A	B
1	Count of Con	Sex
2	Condition	0
3		0	0.00
4		1	

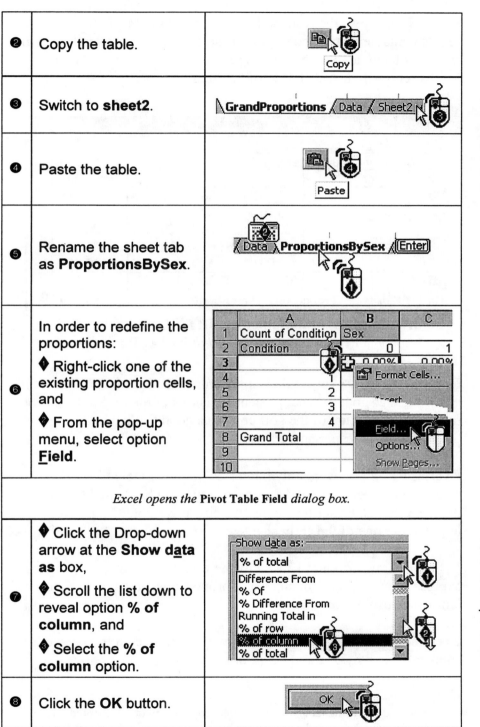

❷	Copy the table.	
❸	Switch to **sheet2**.	
❹	Paste the table.	
❺	Rename the sheet tab as **ProportionsBySex**.	
❻	In order to redefine the proportions: ◆ Right-click one of the existing proportion cells, and ◆ From the pop-up menu, select option **Field**.	

Excel opens the **Pivot Table Field** *dialog box.*

❼	◆ Click the Drop-down arrow at the **Show data as** box, ◆ Scroll the list down to reveal option **% of column**, and ◆ Select the **% of column** option.	
❽	Click the **OK** button.	

Note:
By copying the original pivot table to another sheet, you can freely experiment with the copy without affecting the original.

Note:
If you wanted to generate proportions calculated separately for each **Condition** *category, you would need to select the* **% of row** *option.*

Figure 12.4 shows the results. For example, **11.43%** of all male respondents ranked the condition of the golf course as excellent. On the other hand, among the female respondents, that proportion is **7.14%** (compare it with **1.49%** of the total).

	A	B	C	D	E
1	Count of Condition	Sex			
2	Condition	0	1	2	Grand Total
3	0	0.00%	0.00%	7.14%	1.49%
4	1	0.00%	0.00%	3.57%	0.75%
5	2	100.00%	24.76%	28.57%	26.12%
6	3	0.00%	63.81%	53.57%	61.19%
7	4	0.00%	11.43%	7.14%	10.45%
8	Grand Total	100.00%	100.00%	100.00%	100.00%

Data \ **ProportionsBySex** \ Sh

Figure 12.4 Cross-tabulation Result for Column Total Proportions of *Condition* by *Sex* Responses

Exercise 1. How many male respondents ranked the condition of the golf course as good? What percentage is this of the respondents who ranked the condition of the golf course as good? What percentage is this of the male respondents?

Exercise 2. If you were only interested in the variable *Condition*, what other Excel command sequence could you have used to obtain the frequency distribution for just the single variable? Hint: you learned it in one of the early chapters.

Figures 12.3 and 12.4 show the *Condition* by *Sex* frequencies expressed as percents. In analyzing proportions, you also need to know the absolute counts.

Exercise 3. Create a pivot table that will absolute counts of responses, cross-tabulated by *Sex* and *Condition*.

Note:
To copy the table, click
A1 *(twice, if needed).*
Click the **Copy** *button.*
Select (click) cell **A10**.
Click the **Paste** *button.*

Hint:
Reuse one of the existing tables. For example, copy the pivot table stored on the sheet **ProportionsBySex** and paste it right below, on the same sheet, starting at cell **A10**. As you did it before (Procedure 12.2, step ➏), place the mouse pointer on one of the new data cells (e.g. **B12**) and right-click the mouse. From the pop-up menu,

select option **Field**. From the **Show data as** list, select option **Normal**, and then okay this operation. If necessary adjust the width of columns **B-E**. Figure 12.5 shows both the percent and absolute counts.

	A	B	C	D	E
1	Count of Condition	Sex			
2	Condition	0	1	2	Grand Total
3	0	0.00%	0.00%	7.14%	1.49%
4	1	0.00%	0.00%	3.57%	0.75%
5	2	100.00%	24.76%	28.57%	26.12%
6	3	0.00%	63.81%	53.57%	61.19%
7	4	0.00%	11.43%	7.14%	10.45%
8	Grand Total	100.00%	100.00%	100.00%	100.00%
9					
10	Count of Condition	Sex			
11	Condition	0	1	2	Grand Total
12	0			2	2
13	1			1	1
14	2	1	26	8	35
15	3		67	15	82
16	4		12	2	14
17	Grand Total	1	105	28	134
18					

|◄ ◄ ► ►|\ Data \ **ProportionsBySex** \ Sh |◄ | ⟩|

**Figure 12.5 Column Proportions and Absolute
Frequencies for Cross-tabulation of
the Variables Condition and Sex**

Exercise 4. Based on one of the existing tables, develop a pivot table, to show the proportions as percents of rows. Then explain how the excellent ranking for the condition of the golf course is split between men and women. Store the table on a new sheet, say **ProportionsByCondition**.

Section 12.5 Hypothesis Test on a Single Proportion Using Excel

The next step is to use Excel to run a hypothesis test on a single proportion. Unfortunately Excel does not have a built in test for proportions as it does for means. It can, however standardize any value by subtracting a mean and dividing the difference by a standard deviation. This is precisely what we need since the test statistic for a test on a single proportion is a **Z** statistic:

$$Z = \frac{\hat{p} - p}{\sqrt{p(1 - p)/n}}$$

where: $\hat{p}$ is the sample proportion from the data,

  p is the value you are testing it against,

and n is the sample size.

The denominator of this expression is the standard error of the estimate, $\hat{p}$. In order to get Excel to calculate the **Z** test statistic, you need to specify three values:

1 The sample proportion
2 The hypothesized proportion
3 The sample size

Both **1** and **3** can be obtained from the pivot tables which you have created. Let's continue to examine the variable *Condition*.

Suppose the Board of Directors wishes to advertise that more than half of their members feel that the condition of the golf course is "good" (coded **3**). From the cross tabulation table shown in Figure 12.5, you can see that **82** members rated the condition of the golf course as "good". This is the first step in obtaining the sample proportion to be used in the hypothesis test. The sample proportion would be **82/134 = 0.6119** (see Figure 12.5), if there were no non-response.

Section 12.5.1 Handling the Item Non-Response

Notice that there were two respondents who did not rank the condition of the golf course. These are coded zero (**0**). Because of this, the proportions should be calculated out of **132** responses instead of out of **134** responses to this question. In order to make this adjustment to our pivot tables on the sheet **ProportionsBySex**, you will use the **PivotTable Wizard** to ignore the zero **Condition** responses.

Procedure 12.3 Hiding Data in A Pivot Table

	Task Description	Mouse/Keyboard
❶	Double-click cell **A11**.	![table with cells: A9, A10 Count of Condition, B Sex, A11 Condition, A12, A13]
	Excel opens the **Pivot Table Field** *dialog box.*	
❷	From the **Hide items** list, select option **0**.	Hide items: 0 1 2 3 4
❸	Click the **OK** button.	OK

Now, repeat this operations with the percent table (first double-click the **Condition** field in cell **A2**). Figure 12.6 shows the modified tables.

It is a good time to save your workbook.

	A	B	C	D	E
1	Count of Condition	Sex			
2	Condition	0	1	2	Grand Total
3	1	0.00%	0.00%	3.85%	0.76%
4	2	100.00%	24.76%	30.77%	26.52%
5	3	0.00%	63.81%	57.69%	62.12%
6	4	0.00%	11.43%	7.69%	10.61%
7	Grand Total	100.00%	100.00%	100.00%	100.00%
8					
9					
10	Count of Condition	Sex			
11	Condition	0	1	2	Grand Total
12	1			1	1
13	2	1	26	8	35
14	3		67	15	82
15	4		12	2	14
16	Grand Total	1	105	26	132

◄◄ ► ►◄ / Data \ **ProportionsBySex** / Sh ◄ ►

**Figure 12.6 Modified Column Proportions and
Absolute Frequencies for Cross-
tabulation of Condition and Sex**

Continuing with our example, the sample proportion of the "good" rating is found to be **82/132 = 0.6212**. Thus $\hat{p} = 0.6212$ and **n = 132**.

Section 12.5.2 Using Excel to find the Test Statistic

Knowing the formulas for the **Z** statistic and the Normal distribution functions, you could construct the single proportion test all by yourself. Nevertheless, you may wish to settle for a quick solution via the **SinglePropTest** macro.

Procedure 12.4 Generating a Single Proportion Test

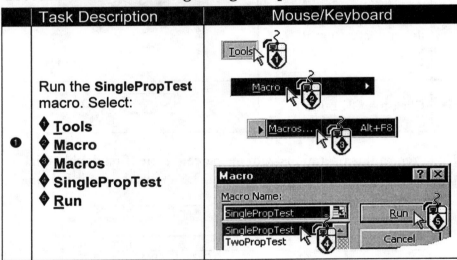

	Task Description	Mouse/Keyboard
❶	Run the **SinglePropTest** macro. Select: ◆ **Tools** ◆ **Macro** ◆ **Macros** ◆ **SinglePropTest** ◆ **Run**	

The macro produces the following output:

	A	B	C	D	E	F
1						
2		Test Statistic for Proportion				
3						
4		**Sample Proportion**	p_s	0.5		
5		**Hypothesized Proportion**	p	0.5		
6		**Sample Size**	n	100		
7						
8		**Test Statistic**	Z	0.0000		
9		**Significance Level**	α	0.05		
10		*Critical Values*				
11		Case I	Z1	-1.6449	p-value1	0.5000
12		Case II	Z2	1.6449	p-value2	0.5000
13		Case III	Z3	-1.9600		
14						

Sheet1 / ProportionsBySex / Sheet3

❷	In the **Sample Proportion** cell (**D4**), enter **0.6212**, or copy this value from cell **E5** of the sheet **ProportionBySex**.	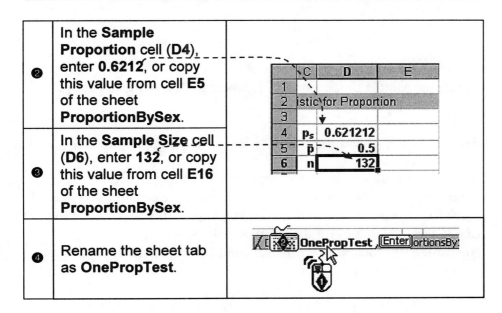
❸	In the **Sample Size** cell (**D6**), enter **132**, or copy this value from cell **E16** of the sheet **ProportionBySex**.	
❹	Rename the sheet tab as **OnePropTest**.	

Figure 12.7 shows the single proportion test outcomes.

Figure 12.7 The Z Test Statistic Model for a Single Proportion

For this example, the **Z** statistic is **2.7852**. It is produced by the formula in cell **D8** (=(D4-D5)/SQRT(D5*(1-D5)/D6)) which is the Excel implementation of the **Z** statistic:

$$Z = \frac{\hat{p} - p}{\sqrt{p(1-p)/n}}$$

Notice the sample proportion, $\hat{p}$, is labeled in the worksheet as p_s.

Section 12.5.3 Deciding to Fail to Reject H_0 or to Reject H_0

Note:
*Use the **critical Z** value(s) to decide if you should reject the null hypothesis.*

There are two ways to decide, if you should reject the null hypothesis (H_0). One way is to compare the **Z** statistic to the critical **Z** value for a given value of alpha, α. In Chapter 10 you learned how to find the critical values for the Normal distribution using the **NormSInv( )** function (the inverse, cumulative, standard, normal probability function).

There are three different cases, which you may encounter when doing a hypothesis test on proportions. The critical values for each case are found using the approach you learned in Chapter 10. The only difference is the value, which you supply as **probability.**

The first two cases are shown below. In these cases you are doing a one sided test on proportions. The specific value of **p** being tested is **0.50.**

Case I:	H_0:	$p \geq 0.50$	Case II:	H_0:	$p \leq 0.50$
	H_A:	$p < 0.50$		H_A:	$p > 0.50$

Lower-Tail Test
Reject H_0 if the test statistic falls below the critical value.

For Case I, you would reject H_0, if the **Z** statistic is "too small" or less than the negative one-sided critical value. This critical value is found by entering the value of alpha (α) being used for the test as **probability.** If alpha is **0.05** you should verify that the critical value is **-1.64485**.

Upper-Tail Test
Reject H_0 if the test statistic falls beyond the critical value.

For Case II, you would reject H_0 if the **Z** statistic is "too large" or greater than the positive one-sided critical value. This critical value is the same as that found for Case I only with the sign changed to a positive number. If alpha is **0.05** you should verify that the critical value is **1.64485**.

The third case is a two-sided test on proportions. A two-sided test against the true value of **0.50** is shown below.

Case III: H_0:   **p = 0.50**
                                 H_A: p $\neq$ 0.50

For Case III, you would reject H_0 if the **Z** statistic were smaller than the negative two-sided critical value or larger than the positive two-sided critical value. The negative two-sided critical value is obtained by entering $\alpha/2$ as **probability**. The positive two-sided critical value is obtained by changing the sign. You should verify that the lower and upper critical values are **-1.96** and **1.96** when $\alpha = 0.05$.

Two-Sided Test
Reject H_0 if the test statistic is less than the lower critical value or greater than the upper critical value.

To determine whether you should use a one-sided test or a two-sided test you need to consider what question you are trying to answer. In our example, since the management would like to say that more than **50%** of their members ranked the condition of the greens as "good" then you would use Case II of the one-sided test. Since the **Z**-statistic is **2.7850**, you would reject the null hypothesis and thus accept the alternative hypothesis (**p > 0.50**).

You can alter the critical regions by changing the value for alpha from **0.05**. Thus, your decision to reject the null hypothesis could be different for different values of alpha.

Exercise 5: Change alpha to **0.01**. What happens to the critical values? Do you reject H_0? Now change alpha to **0.10**. What happens to the critical values? Do you reject H_0?

The second method of deciding whether to reject or fail to reject H_0 is to use the **p-value**. Remember that the **p-value** for a test of a hypothesis is the probability of obtaining a value of **Z** as extreme or more extreme than the sample value when H_0 is true. The **p-value** is also called the observed significance level of the test since it represents the smallest value of alpha for which we could reject H_0 using the observed data. For example, suppose the **p-value** was **0.07**. This means that if you set alpha at **0.05** you would fail to reject H_0 but if you set alpha at **0.10** then you would reject H_0. In fact the decision switches at $\alpha = 0.07$. If the **p-value** is smaller than alpha then you reject H_0. By providing management with the **p-value** they can then see for what value of alpha the decision changes.

*Use the **p-value** to decide if you should reject the null hypothesis.*

Remember the p-value tells you at what value of alpha the decision switches.

The **p-value** can be found by using the cumulative distribution for the Normal distribution, which you learned about in Chapter 7. For our example, the **Z** test statistic is **2.7850**. The **p-value** shown in Figure 12.7 is interpreted as the probability of obtaining **Z** greater than the test statistic calculated from the sample when H_0 is true. For our example (Case II), this probability is approximately **P(Z > 2.7850) = 0.0027 (or 0.27%)**.

One of the important things to remember when conducting any hypothesis test is that as you decrease the probability of a Type I error, α, the probability of a Type II error, β, increases. Thus, it is not always desirable to set α as small as possible because in doing so the probability of a Type II error increases dramatically. By reporting the **p-value**, the decision to reject the null hypothesis (with the potential for a type I or a type II error) is left up to the decision maker.

Exercise 6:   What is the **p-value** for the two-sided test hypothesis test on the variable *Condition*? Is it legitimate for the Board to make the statement they wish regarding how their members feel about the condition of the course?

Section 12.6   Comparing Two Proportions: Hypothesized Difference of Zero

Section 12.6.1 Setting up the Hypothesis Test

In many cases it is desirable not only to look at a single proportion but to compare proportions. For example, the Board of Directors may wish to compare the responses of the young members to the older members. They may wish to compare the responses of the newer members to the members who have been with the Club many years. Another possibility would be to compare the responses of the male members to the female members. Perhaps the married members have different concerns than the single members. All of these comparisons can be examined by running a hypothesis test to compare two proportions.

In this section you will examine whether the locker rooms are rated the same by the men and the women. In particular we will compare the proportion of males who rated the locker rooms "good" to the proportion of females who rated the locker rooms "good". Call the male respondents **population 1** and the female respondents **population 2**. The proportions calculated for **population 1** are then labeled p_1 and those from **population 2** are labeled p_2. The comparison hypothesis test can take one of three forms:

Case I: $H_0 : p_1 - p_2 \le 0$ **Case II:** $H_0 : p_1 - p_2 \ge 0$
        $H_A : p_1 - p_2 > 0$ $H_A : p_1 - p_2 < 0$

Case III: $H_0 : p_1 - p_2 = 0$
           $H_A : p_1 - p_2 \ne 0$

In order to decide which of these three tests you should use, you need to consider what type of information you desire. Most of the time Case III is used to examine whether or not the two groups behave *differently*. For the example use Case III to see if the men rate the locker rooms differently from the women.

Section 12.6.2 Comparing Two Population Proportions in Excel

You most likely learned that the appropriate test statistic for comparing two population proportions is a **Z** test statistic which looks like this:

$$z = \frac{(\hat{p}_1 - \hat{p}_2) - 0}{\sqrt{\dfrac{\bar{p}(1-\bar{p})}{n_1} + \dfrac{\bar{p}(1-\bar{p})}{n_2}}}$$

where $\hat{p}_1$ and $\hat{p}_2$ are the sample proportions and $\bar{p}$ is the pooled sample proportion found by $\dfrac{n'_1 + n'_2}{n_1 + n_2}$. The values for n'_1 and n'_2 are the number in the first and second sample, respectively, with the characteristic of interest. For the example, that is the number of men who rated the condition of the locker rooms as "good" (n'_1) and the number of women who rated the condition of the locker rooms as "good" (n'_2). The pooled proportion is an estimate of the true proportion of members who would rate the locker rooms as "good" if H_O were true.

Unfortunately Excel does not provide a direct test of two proportions based on the **Z** test statistic. However, as it was with the single proportion, Excel is equipped with all necessary functions to develop such a test. This time, you will develop a test model and perform the test to compare proportions p_1 and p_2 associated with the variables *Sex* and *Locker*.

Section 12.6.3   Calculating proportions in Excel

To find out the proportion of males who rated the condition of the locker rooms as "good", we must count the number, n_1', of rows in the data set in which the *Sex* field contains **1** ("male") and *Locker* field contains **3** ("good"). For the proportion of females who rated the condition of the locker rooms as "good", we must count the number, n_2', of rows in the database in which the *Sex* field contains **2** ("female") and *Locker* field contains **3** ("good"). Dividing these results by the respective total number of responding males and females will produce the proportions:

$$\hat{p}_1 = \frac{n_1'}{n_1} \qquad\qquad \hat{p}_2 = \frac{n_2'}{n_2}$$

The **PivotTable Wizard** is a convenient tool when you want to explore all possible cross-tab measures (averages, frequencies, proportions, etc.). Since in this case you are interested in only one *Locker* value ("good"=3) for males (*Sex* = 1) and females (*Sex*=2), you can get all necessary information by means of the **DCount( )** function. The following procedure show all necessary steps.

Procedure 12.5 Calculating Frequencies Using the DCount( ) Function

	Task Description	Mouse/Keyboard
❶	Switch to the next available sheet and rename it as **LockerRating**.	onsBy⬚❎ **LockerRatings** ⏎Enter

		A	B	C	D
❷	Enter labels and number as shown on the right.	1	Males		Females	
		2	Sex	Locker	Sex	Locker
		3	1	3	2	3
		4				
		5	n'1		n'2	
		6	n1		n2	

Note:
The cells A5 and C5 contain labels n'1 and n'2.

❸	Apply formatting approximately as shown on the right. **Use Center-Across Selection** alignment for **A1:B1** and **C1:D1**. Right-align **A2:D2**, **A5:A6**, and **C5:C6**. To show the digits in **A5:A6** and **C5:C6** as subscripts, select each digit on the formula bar and run **Format	Cells	Subscript	OK**.

	A	B	C	D
1		Males		Females
2		Sex	Locker	Sex	Locker
3		1	3	2	3
4					
5		n'_1		n'_2	
6		n_1		n_2	

Note:
The ranges A2:B3 and C2:D3 define the filter criteria for the counts.

Note:
The entire data set on the sheet Data is named as CountryClubSurvey. All its variables are named using the titles stored in the top row cells.

❹	Calculate the number of males that rated the lockers as "good". In B5, enter this formula:

=DCOUNT(CountryClubSurvey,"Locker",A2:B3)

❺	Calculate the number of males. In B6, enter this formula:

=DCOUNT(CountryClubSurvey,"Locker",A2:A3)

❻	Obtain similar counts for females: ◆ Select range **B5:B6**, ❷ Click the **Copy** button, ◆❸ Select cell **D5**, and ◆ Click the **Paste** button.

	A	B	C	D
5		n'_1	29	n'_2	
6		n_1	105	n_2	

Figure 12.8 shows the counts. Cells **B5** and **D5** contain the counts of males (n'_1) and females (n'_2) who rated the condition of the locker rooms as "good", respectively. Cells **B6** and **D6** contain the total numbers of responding males (n_1) and females (n_2), respectively.

	A	B	C	D
1		Males		Females
2		Sex	Locker	Sex	Locker
3		1	3	2	3
4					
5		n'_1	29	n'_2	19
6		n_1	105	n_2	28

LockerRatings / Sheet4 / Sheet5 / St

Figure 12.8 Locker vs. Sex Counts

Thus the proportions are:

$$\hat{p}_1 = \frac{n_1'}{n_1} = \frac{29}{105} = 0.2762 \quad \hat{p}_2 = \frac{n_2'}{n_2} = \frac{19}{28} = 0.6786$$

Exercise 7. Repeat the above procedure with the variable *Greens* to allow for the comparison of the proportion of men and women who feel that the condition of the greens is "poor".

Section 12.6.4   Running the Hypothesis Test for a Comparison

Once you have the sample proportions, you are ready to use Excel to run the hypothesis test.

Procedure 12.6 Generating a Test for Two Proportions

	Task Description	Mouse/Keyboard
①	Run the **TwoPropTest** macro. Select: ♦ **Tools** ♦ **Macro** ♦ **Macros** ♦ **TwoPropTest** ♦ **Run**	
	The macro produces a two-proportion test setup on a new sheet.	
②	In the **Sample 1 Proportion** cell (**C4**), enter **=29/105**, or formula: <div align="center">**=LockerRatings!B5/LockerRatings!B6**</div>	
③	In the **Sample Size, n1** cell (**E4**), enter **=105**, or formula: <div align="center">**=LockerRatings!B6**</div>	
④	In the **Sample 2 Proportion** cell (**C5**), enter **=19/28**, or formula: **=LockerRatings!D5/LockerRatings!D6**	

❹	In the **Sample Size, n2** cell (**E5**), enter **=28**, or formula: **=LockerRatings!D6**
❺	Rename the sheet tab as **TwoPropTest**.	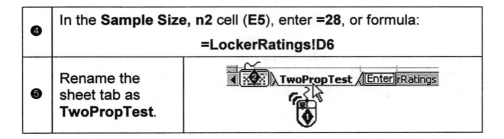

Figure 12.9 shows the two-proportion test outcomes.

	A	B	C	D	E
1	**Test Statistic for Two Proportions**				
2					
3		**Proportion**		**Sample Size**	
4	Sample 1	p1	0.2762	n1	105
5	Sample 2	p2	0.6786	n2	28
6	Pooled	pp	0.3609		
7					
8	Test Statistic	Z	-3.9392		
9	Significance Level	α	0.05		
10	*Critical Values*				
11	Case I	Z1	-1.6449	p-value1	0.000041
12	Case II	Z2	1.6449	p-value2	0.999959
13	Case III	Z3	-1.9600		

TwoPropTest / LockerRatings / Sheet4 / Sh

**Figure 12.9 Two-proportion Test Outcomes for *Locker
Rating by *Sex***

Note that interpretation of the two-proportion test is similar to the test for a single proportion. To apply this test to other proportions, simply enter new proportions into cells **C4**, **C5**, and new sample sizes into cells **E4**, **E5**.

*It is a good time to
save your workbook.*

Exercise 8. Using the output shown in Figure 12.9, what can you conclude about the proportion of men (p_1) and women (p_2) who feel that locker rooms are in good condition? What suggestions might you have for the Board of Directors?

Section 12.7 Investigative Exercises

1. Describe the respondents (demographics) of the survey of the Country Club members. Use whatever graphs and descriptive statistics you feel are appropriate.

2. The response rate for this survey was **53.6%** which is high for a mail survey. Even though there was a high response rate, the results could still be biased by the non-response. Bias creeps in if the non-respondents "look" different from the respondents. You should try to get some of the non-respondents to respond by using a second mailing or a phone interview. Speculate about the type of member who would be inclined to not respond. What would you recommend to the Board of Directors?

3. Complete the following table with the appropriate proportions:

Rated the:	Exc.	Good	Fair	Poor
Condition of the Golf course				
Condition of the greens				
Condition of the landscape				
Parking accessibility				
Condition of the locker rooms				

What areas seem to be in need of improvement?

General Instructions for Exercises 4-11: No alpha value is specified for the hypothesis tests in exercises 4-7. You should report the **p-values** and explain what the **p-value** means in each case.

---

4. Examine the proportion of respondents who rated the condition of the *Greens* "good". Conduct the following hypothesis test:

$$H_o: \quad p \geq .50$$
$$H_A: \quad p < .50$$

What is your conclusion about the true proportion of members who feel the condition of the *Greens* is "good"?

---

5. What proportion of the respondents rated the *Landscape* "fair" or "poor"? Does this mean that you would automatically reject the null hypothesis in the following test:

$$H_o: \quad p \leq .20$$
$$H_A: \quad p > .20$$

Conduct the hypothesis test to verify your answer.

---

6. Can you conclude that less than 5% feel that the *Parking* accessibility is "poor"?

---

7. Can you conclude that more than 50% feel that the *Condition of the locker rooms* is "poor"?

8. Is there any significant difference in the rating of the *condition* of the golf course by sex? by age? by income? by marital status? Examine only the "fair" rating. What are your conclusions about the members who consider the condition of the golf course "fair"? Consider using the following table to display your results:

Sex	Proportion rated 'fair'	Age	Proportion rated 'fair'
Male		40 or Under	
Female		Over 40	
Significantly Different?		Significantly Different?	

Income	Proportion rated 'fair'	Marital Status	Proportion rated 'fair'
Under $50,000		Single	
Over $50,000		Married	
Significantly Different?		Significantly Different?	

9. Is there any significant difference in the rating of the condition of the *greens* by sex? by age? by income? by marital status? Examine only the "fair" rating. What are your conclusions about the members who consider the condition of the greens "fair"?

Sex	Proportion rated 'fair'	Age	Proportion rated 'fair'
Male		40 or Under	
Female		Over 40	
Significantly Different?		Significantly Different?	

Income	Proportion rated 'fair'	Marital Status	Proportion rated 'fair'
Under $50,000		Single	
Over $50,000		Married	
Significantly Different?		Significantly Different?	

10. Is there any significant difference in the rating of the surrounding *landscape* of the golf course by sex? by age? by income? by marital status? Examine only the "fair" rating. What are your conclusions about the members who consider the landscape "fair"?

Sex	Proportion rated 'fair'	Age	Proportion rated 'fair'
Male		40 or Under	
Female		Over 40	
Significantly Different?		Significantly Different?	

Income	Proportion rated 'fair'	Marital Status	Proportion rated 'fair'
Under $50,000		Single	
Over $50,000		Married	
Significantly Different?		Significantly Different?	

11. Is there any significant difference in the rating of the *parking* accessibility by sex? by age? by income? by marital status? Examine only the "fair" rating. What are your conclusions about the members who consider the parking accessibility "fair"?

Sex	Proportion rated 'fair'	Age	Proportion rated 'fair'
Male		40 or Under	
Female		Over 40	
Significantly Different?		Significantly Different?	

Income	Proportion rated 'fair'	Marital Status	Proportion rated 'fair'
Under $50,000		Single	
Over $50,000		Married	
Significantly Different?		Significantly Different?	

12. On the basis of your analysis, what areas do you feel need improvement? What recommendations do you have for increasing membership for this Country Club?

Chapter 13 "Country Club Members: A Closer Look"

Chi-Square Test for Independence

Section 13.1 Overview

Statistical Objectives: After reading this chapter and doing the exercises you will:

- Know when it is appropriate to use a Chi-Square test for independence.
- Know how to interpret the results of the tests.

Section 13.2 Problem Statement

In Chapter 12 you looked at the results of a survey of the members of a New England country club. In particular, you examined a variety of questions related to member satisfaction. In that chapter you examined single proportions and compared two proportions. You were able to draw some conclusions based on this analysis. In this chapter you will continue to analyze this survey data.   For example, you may wish to know whether members in several age groups have the same opinion about the condition of the golf course. You may also wish to know if a member's income is related to his/her rating of the landscape surrounding the golf course.

Open the data set named **Ch13Dat.xls** from your data disk. Note that Procedure 3.1, on page 33, contains detail instruction about how to open an Excel workbook.

Click the Open *button or use the* File | Open *command (*ALT*+*F*,* O*) to open the file.*

Section 13.3 Chi-Square Test for Independence

In order to explore questions similar to the ones suggested in the previous section, you need to use a Chi-Square test for independence. Two criteria of classification are said to be *independent* if the distribution of one criterion in no way depends on the distribution of the other. For example, consider the variables *Age* and *Condition*. You wish to be able to tell the Board of Directors if the different age groupings rated the condition of the golf course the same way. If they are not the same then at least one of the age groupings may have some special needs or concerns, which are not being addressed.

Look at the cross tabulation table of *Age* by *Condition* shown in Figure 13.1.

Note:
Procedure 12.1 (page 287) shows detail steps for doing cross-tabulation. The **Pivot Table Wizard - Step 3 of 4** *should result in:*

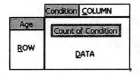

where **Count of Condition** *is a straight count (not percent).*

	A	B	C	D	E	F	G
1	Count of Condition	Condition					
2	Age	0	1	2	3	4	Grand Total
3	0				1	1	1	3
4	1			2	1	1	4
5	2			8	9		17
6	3		1	7	32	8	48
7	4	2		17	39	4	62
8	Grand Total	2	1	35	82	14	134

Ch13Dat.xls — ConditionByAge / Data / Sheet2 / Sheet3

Figure 13.1 Observed Frequencies of Condition by Age

You can see that **1** of the members under **20** (row variable coded **1**) rated the condition of the golf course as "excellent" (column variable coded **4**), **1** rated it "good", **2** rated it "fair" and **0** rated it "poor". The question of interest is whether or not the distributions of the ratings are the same for the members aged **21-40**, **41-60** and over **60**. If, in fact, it appears that the age groups rated the condition of the golf course the same way then the variables *Condition* and *Age* are considered independent.

The hypothesis you are testing is:

H₀: *Age* and rating of *Condition* of the Golf Course are independent

Hₐ: *Age* and rating of *Condition* of the Golf Course are not independent

The following steps should be followed in a Chi-Square test of independence.

■ From the populations of interest, draw random samples.

■ Cross-tabulate the sample. The levels of one variable provide row headings and the levels of the other variable provide column headings. The cell entries in this table are called observed frequencies. Such a table is also called a contingency table.

■ Compute the χ^2 statistic and compare it to the upper critical value.

■ If the computed χ^2 statistic is equal to or greater than the upper critical χ^2, then reject H_0 and conclude that the variables are not independent. Otherwise do not reject H_0 and conclude that the variables are independent.

Note:
Remember that, if two variables are independent, it means that how observations are categorized on one variable has no bearing on how they are categorized on the other variable.

The first two steps are almost taken care of already. The samples are collected. Figure 13.1 shows the observed (actual) cross-tab frequencies. Notice, however, that the frequencies and the row/column totals take into consideration also the non-response data (coded as zeros). Before moving to the next step, you must modify the table so it does not reveal the non-response outcomes.

To eliminate (hide) the non-response data for the variable *Age*, double-click the **Age** button. Excel opens the **PivotTable Field** dialog box. From the **Hide Items** list, choose **0** and okay this operation. In a similar way hide the non-response data for the variable *Condition*. Your pivot table should now look as the one in Figure 13.2.

	A
1	Count of Condition
2	Age
3	0
4	1

Ch13Dat.xls

	A	B	C	D	E	F
1	Count of Condition	Condition				
2	Age	1	2	3	4	Grand Total
3	1		2	1	1	4
4	2		8	9		17
5	3	1	7	32	8	48
6	4		17	39	4	60
7	Grand Total	1	34	81	13	129

ConditionByAge / Data / Sheet2 / She

**Figure 13.2 Observed Frequencies of Condition by Age
Without Non-response Data**

Another modification of the pivot table is needed before we can apply the Chi-square test. One of the Excel functions that will be used to perform the test requires that there be no empty cells in the ranges containing the actual and expected frequencies. If you check Figure 13.2, you will see a few empty cells (B3:B4, B6, and E4). The following procedure shows how to replace the empty cells with zeros.

Procedure 13.1 Replacing Empty Cells of a Pivot Table with Zeros

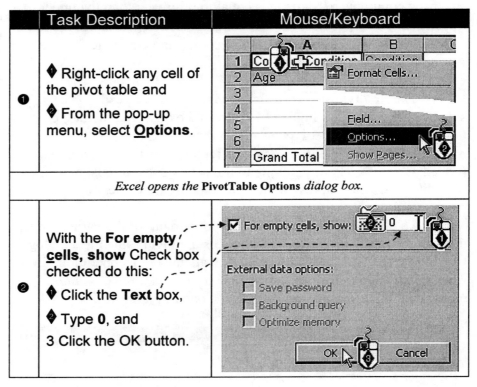

Task Description	Mouse/Keyboard
❶ ◆ Right-click any cell of the pivot table and ❷ From the pop-up menu, select **Options**.	
Excel opens the **PivotTable Options** *dialog box.*	
❷ With the **For empty cells, show** Check box checked do this: ◆ Click the **Text** box, ❷ Type **0**, and 3 Click the OK button.	

In order to calculate the χ^2 statistic, you also need to produce the expected frequencies, which are defined as follows:

$$ExpectedFrequency = \frac{RowSum * ColumnSum}{TotalSum} = \frac{AgeCount * ConditionCount}{TotalCount}$$

Notice that the **PivotTable Wizard** has already calculated the row, column, and total sums. To produce the expected frequencies do this:

Procedure 13.2 Calculating the Expected Frequencies for Condition by Age Responses

	Task Description	Mouse/Keyboard
❶	On the **ConditionByAge** sheet, go to cell **B8** and enter label **Expected Frequencies**.	[F5] B9 [Enter] Expected Frequencies [Enter]
❷	Center the label across columns **B:E**.	Merge and Center
❸	Change the font style to **Bold**.	Bold
❹	Apply the **Bold** font style and fill the range with a light **Green** color.	No Fill — Light Green
❺	Select range **B10:E13**.	
❻	Type formula **=B$7*$F3/$F$7** and press [CTRL]+[ENTER].	

❼	Round the results to **0.01**.	

Figure 13.3 shows the expected frequencies.

	B	C	D	E
9	**Expected Frequencies**			
10	0.03	1.05	2.51	0.40
11	0.13	4.48	10.67	1.71
12	0.37	12.65	30.14	4.84
13	0.47	15.81	37.67	6.05

Figure 13.3 Expected Frequencies of Condition by Age

To obtain the **Chi-Square** test related data, perform the following operations.

Procedure 13.3 Doing the Chi-square Test for Independence

	Task Description	Mouse/Keyboard
❶	Enter labels and numbers as shown on the right.	<table><tr><td></td><td>A</td><td>B</td></tr><tr><td>15</td><td>α</td><td>5%</td></tr><tr><td>16</td><td>df</td><td>9</td></tr><tr><td>17</td><td>p-value</td><td></td></tr><tr><td>18</td><td>Chi-Sq Statistic</td><td></td></tr><tr><td>19</td><td>Chi-Sq Critical</td><td></td></tr></table>
❷	For the **p-value**, move to cell **B17** and enter the ChiTest( ) function with references to the actual and expected frequencies.	[F5] **B17** [Enter] =CHITEST(B3:E6,B10:E13) [Enter]

Note:
To obtain a Greek *letter* α, *enter a Latin letter* a *into cell* **A15** *and change its font face to* **Symbol**.

❸	To calculate the **Chi-Sq Statistic**, go to cell **B18** and enter the Chi-square inverse function, **ChiInv( )**, with references to the **p-value** and degrees of freedom (**df**).	[F5] B18 [Enter] =CHIINV(B17,B16) [Enter]
❹	Finally, to get the **Chi-Sq Critical** value, in cell **B19**, enter the Chi-square inverse function with references to the significance level (α) and degrees of freedom (**df**).	[F5] B19 [Enter] =CHIINV(B15,B16) [Enter]

Note:
*The degrees of freedom value (**df**) is based on the number of rows and columns of the contingency table:*

df = (NumberOfRows -1)* (NumberOfColumns -1)

Figure 13.4 shows the results.

B19	▼	=	=CHIINV(B15,B16)

	A	B	C	D
15	α	5%		
16	df	9		
17	p-value	0.102369		
18	Chi-Sq Statistic	14.60511		
19	Chi-Sq Critical	16.91896		

Figure 13.4 χ^2 **Test for Independence Results**

The **Chi-Square** statistic is the sum of the squared-relative differences between the actual and expected frequencies:

$$\chi_0^2 = \sum_{i=1}^{m} \sum_{j=1}^{n} \frac{\left(f_{i,j}^a - f_{i,j}^e\right)^2}{f_{i,j}^e}$$

Instead of χ_0^2, Excel uses the ChiTest() function to produce a **p-value** associated with χ_0^2. Thus, you can obtain the **Chi-Square** statistic, χ_0^2, using the inverse of the **Chi-Square** distribution function, $\chi_0^2 =$ ChiInv(**p-value**,**df**).

Note:
Using the p-value we would arrive at the same conclusion of failing to reject Ho, since p-value > α .

By examining the output shown in Figure 13.4, you can see that the **Chi-Square** statistic is **14.605**. You would reject H_O, if the **Chi-Square** statistic was larger than the critical value which is **16.919** (for $\alpha = 0.05$). Thus, we fail to reject H_O since the calculated test statistic value is not greater than the critical value. We conclude that *Age* and *Condition* are independent variables.

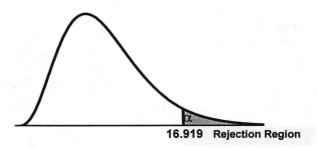

Figure 13.5 χ^2 Distribution and the Critical Value (for $\alpha=5\%$, df=9)

You should remember that one of the conditions for using the **Chi-Square** test is that there be a minimum of **5** observations in each cell of the contingency table. This condition is violated in our example of *Age* and *Condition*. This means that there were too few (**<5**) observations in some of the cells and you should collapse some of the categories and re-run the test. In order to do this you may have to combine two or more rows or columns together in the observed frequencies.

Section 13.4 Investigative Exercises

General Instructions for Exercises 1-8: No alpha value is specified for the hypothesis tests in exercises 1-8. You should report the **p-**values and explain what the **p-**value means in each case.

1. Investigate the member rankings of the *Condition* of the golf course. Is the *Condition* of the golf course dependent on the members *Age*, *Sex*, or *Income*? What recommendations do you have based on this information?

2. Investigate the member rankings of the condition of the *Greens*. Is the condition of the *Greens* dependent on the members *Age*, *Sex*, or *Income*? What recommendations do you have based on this information?

3. Investigate the member rankings of the condition of the *Landscape* surrounding the golf course. Is the condition of the *Landscape* dependent on the members *Age*, *Sex*, or *Income*? What recommendations do you have based on this information?

4. Investigate the member rankings of the *Parking* accessibility. Is the *Parking* accessibility dependent on the members *Age*, *Sex*, or *Income*? What recommendations do you have based on this information?

5. Investigate the member rankings of the condition of the *Locker* room. Is the condition of the *Locker* room dependent on the members *Age*, *Sex*, or *Income*? What recommendations do you have based on this information?

6. Is the variable *Howlong* dependent on the variable *Often*? What does your analysis tell you about the membership at this country club?

7. Is the variable *Type* related to the members *Age* ? Is this surprising?

8. Conduct tests for independence on any other variables that you feel should be examined.

9. On the basis of your analysis, what recommendations would you make to the Board of Directors of this country club? Your recommendations should identify any particular groups who might have special needs and/or concerns, suggest ways to increase their membership and identify areas in need of improvement.

Chapter 14 "How Are They Related?"

Linear Regression and Correlation

Section 14.1 Overview

Statistical Objectives: After reading this chapter and doing the exercises a student will:

- Know the difference between a dependent and independent variable and how to determine which is which.
- Know how to find a linear regression equation for a pair of variables.
- Know how to interpret the output of a linear regression to determine whether a significant relationship exists.
- Know what the coefficient of determination is and what it means.
- Know how to use the regression equation to predict values of the dependent variable.
- Know what residuals are and what they mean.
- Know the difference between interpolation and extrapolation.

Section 14.2 Problem Statement

A company that manufactures computer storage media is wondering whether the **Total Quality Management (TQM)** that they have introduced in their floppy diskette manufacturing line is achieving the results they expected.

The program, which incorporates among other things, Quality Circles, Statistical Quality Control and Team Based Decision-Making, has been in place for almost two years. According to the literature when such programs are used they should result in increased productivity and quality and decreased waste and delay. The management of the company asked the production team to assemble some data to assess the success of the program before they introduce similar programs into other areas of the corporation.

Before they collected any data, the team assembled a checklist of what they think they know about the process and its behavior over the past two years. They all agree that they have noticed an increase in the speed at which the production line runs and most people think that there has been a decrease in the amount of waste from the process. They are not sure that the increase in speed means that there has been an increase in productivity, nor are they really sure that there has been a decrease in delay. They are almost all sure that they saw an initial improvement in quality, but that the improvement was not sustained.

They decided to look at five different variables from the production process to see if their initial reactions to the question are correct. They then collect monthly data on Machine Speed, Waste, Delay, Rate of Operation, and Average Outgoing Quality for the two-year period since the TQM program was introduced.

Section 14.3 Characteristics of the Data Set

FILENAME:	Ch14Dat.xls	An Excel Workbook
SIZE:	COLUMNS	6
	ROWS	25

A fragment of the data file is shown in Figure 14.1.

Note:
All the variables are already named using the titles stored in the top row.

	A	B	C	D	E	F
1	Month	Speed	Waste	RateOper	Delay	Quality
2	1	375	8.9	14.5	6.2642	95
3	2	334	9.9	12.8	6.4854	93
4	3	356	8.7	12.7	6.8372	94
5	4	378	9.5	13.9	5.7134	93
6	5	373	9.8	13.7	6.3136	94
7	6	381	8.8	14.7	6.2034	92

Figure 14.1 The TQM Data Set

Notes on the data file:

1. The variable *Month* is the number of months since the TQM program was introduced and is a number from 1 to 24.

2. The variable *Speed* is the number of items per minute that the machine is set to produce.

3. The variable *Waste* is the percent of manufactured product that is discarded during the manufacturing process.

4. The variable *RateOper* is the rate of operation and is a calculated number, which measures the usable product throughput.

5. The variable *Delay* is a measure of the number of minutes per hour that the process is not operating.

6. The variable *Quality* is the percent non-defective product in those product lots that are shipped.

Open the data set named **Ch13Dat.xls** from your data disk. Note that Procedure 3.1, on page 33, contains detail instruction about how to open an Excel workbook.

Note:
To refer to a variable, you can use its spreadsheet address or name. For example, the **Speed** *variable can be referenced via either its name* **Speed** *or spreadsheet address* **Data!B2:B26**.

Open

Click the Open *button or use the* File | Open *command (*ALT*+*F*, *O*) to open the file.*

Section 14.4 Regression and Correlation

While many people speak of regression analysis and correlation analysis as if they are one and the same, they do in fact have very different purposes. **Correlation** analysis is used to determine whether two variables are *related* in a linear manner and provides an estimate of ρ, the correlation coefficient. **Regression** analysis provides a model (function) for predicting or approximating the value of a *dependent* variable using a set of one or more *independent* variables. In the simple case the model produced is linear, that is the equation of a straight line.

The manufacturing team for the floppy diskettes is interested in knowing how different variables in their process are related. Since they want to know *how* as well as *if* they should probably look at the data using regression analysis. Correlation information is usually provided along with a regression analysis, but not vice versa.

Simple linear regression assumes that there is a single, *controllable*, independent variable and a single, dependent variable and that these two variables are related according to the equation:

$$Y = \beta_O + \beta_1 X$$

where:

β_1 is the slope coefficient and

β_O is the y-intercept value.

The Least Squares method is used to find estimates for the values β_O and β_1 and yields and equation of the form:

$$Y = b_0 + b_1 X.$$

Section 14.5 Looking for Relationships

As always, before launching any statistical analysis you need to **look** at the data and assess what is going on. This will make the results of any analysis you do later on much more understandable. Since linear regression looks at the relationship between a controllable, independent variable and tries to find a model to predict values of the dependent variable, the first task is usually to decide which is which!

Looking at the variables in the data set and knowing what the team are interested in finding out, it would appear that the **independent** variable must be *Speed*, since this is the only variable that is within the control of the process. Suppose you look first at the relationship between *Speed* and *Waste*.

In order to determine whether there is even any reason to suspect that *Speed* and *Waste* are related you need to make a Scatter Plot of the data. You can do this in Excel by making an **XY**-Scatter chart for the data. Here **X** is to be represented by *Speed* and **Y** by *Waste*.

Procedure 14.1 Generating an XY Scatter Plot for Waste vs. Speed

Note:
Press [END] *followed by navigation keys while holding down* [SHIFT], *in order to quickly select a large range of cells filled with data.*

	Task Description	Mouse/Keyboard
❶	Copy variables *Speed* and *Waste* and paste them onto **sheet2** at cell **A8**.	[F5] **Data!B1** [Enter] [Shift][End][↓][→] [Ctrl][C] [F5] **Sheet2!A8** [Enter] [Ctrl][V]

❷	While the copies of the variables *Speed* and *Waste* are still selected (highlighted), invoke the **Chart Wizard**.

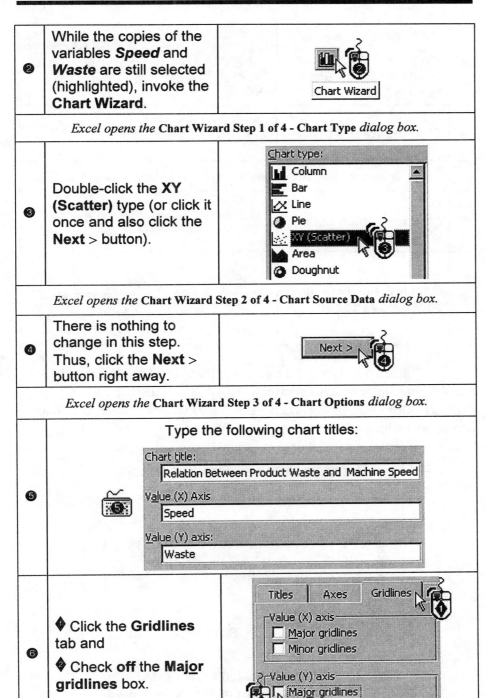

Excel opens the **Chart Wizard Step 1 of 4 - Chart Type** *dialog box.*

❸	Double-click the **XY (Scatter)** type (or click it once and also click the **Next >** button).

Excel opens the **Chart Wizard Step 2 of 4 - Chart Source Data** *dialog box.*

❹	There is nothing to change in this step. Thus, click the **Next >** button right away.

Excel opens the **Chart Wizard Step 3 of 4 - Chart Options** *dialog box.*

❺	Type the following chart titles: Chart title: Relation Between Product Waste and Machine Speed Value (X) Axis Speed Value (Y) axis: Waste

❻	❶ Click the **Gridlines** tab and ❷ Check **off** the **Major gridlines** box.

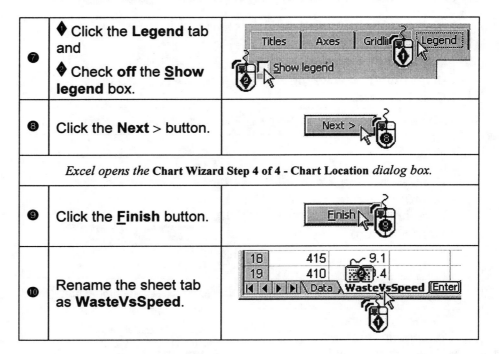

❼	♦ Click the **Legend** tab and ♦ Check **off** the **Show legend** box.
❽	Click the **Next** > button.

Excel opens the **Chart Wizard Step 4 of 4 - Chart Location** *dialog box.*

❾	Click the **Finish** button.
❿	Rename the sheet tab as **WasteVsSpeed**.

Figure 14.2 shows the resulting chart. By default, Excel sets the **XY** coordinate system starting at **(0,0)**. Since there are no **X** (*Speed*) values below **300** and no **Y** (*Waste*) value below **7**, the chart can be modified to reflect those lower limits.

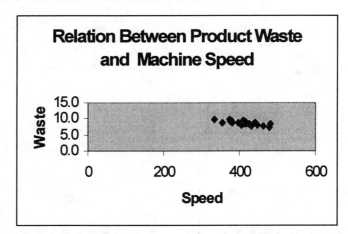

Figure 14.2 Chart Wizard - Generated
Scatter Plot of Waste vs. Speed

In the next few steps, you will modify the chart's **X** and **Y** scopes and change other attributes.

Procedure 14.2 Modifying an XY-Scatter Plot

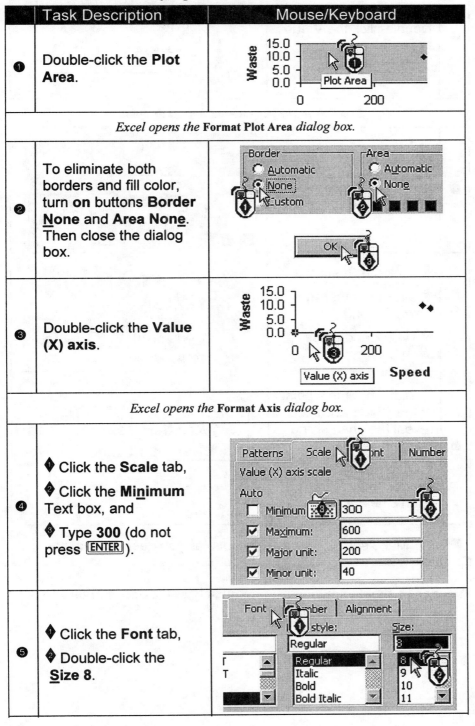

	Task Description	Mouse/Keyboard
❶	Double-click the **Plot Area**.	
	Excel opens the **Format Plot Area** *dialog box.*	
❷	To eliminate both borders and fill color, turn **on** buttons **Border None** and **Area None**. Then close the dialog box.	
❸	Double-click the **Value (X) axis**.	
	Excel opens the **Format Axis** *dialog box.*	
❹	◆ Click the **Scale** tab, ◆ Click the **Minimum** Text box, and ◆ Type **300** (do not press ENTER).	
❺	◆ Click the **Font** tab, ◆ Double-click the **Size 8**.	

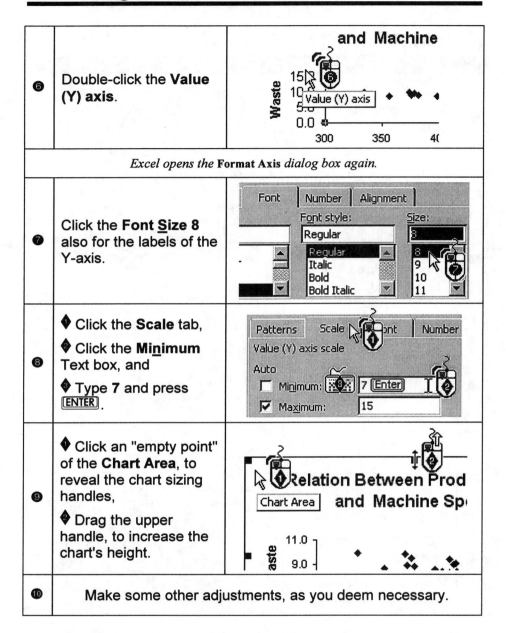

⑥	Double-click the **Value (Y) axis**.	
	Excel opens the **Format Axis** *dialog box again.*	
⑦	Click the **Font Size 8** also for the labels of the Y-axis.	
⑧	◆ Click the **Scale** tab, ◆ Click the **Minimum** Text box, and ◆ Type **7** and press ENTER.	
⑨	◆ Click an "empty point" of the **Chart Area**, to reveal the chart sizing handles, ◆ Drag the upper handle, to increase the chart's height.	
⑩	Make some other adjustments, as you deem necessary.	

Your chart should now look approximately as the one in Figure 14.3.

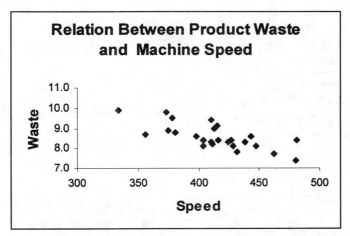

Figure 14.3 Scatter Plot Showing Relationship Between Product Waste and Machine Speed

Based on the chart, it is reasonable to assume that a relationship exists between the two variables and that it is indeed linear.

Section 14.6 Measuring Relationships

The first thing that you might want to do after looking at the graph is to determine the strength of the relationship between *Waste* and *Speed*. You can do this by finding the covariance and/or correlation coefficient for the two variables using the **Covar()** and **Correl()** functions.

Procedure 14.3 Calculating Covariance and Coefficient of Correlation

	Task Description	Mouse/Keyboard
❶	On the **WasteVsSpeed** sheet, enter the following labels: ◆ In **A1**: **Measuring Relationship Between Waste and Speed**, ❷ In **C3**: **Covariance** ◆ In **C4**: **Coefficient of Correlation**.	

Note:
Also change the font style of the labels to **Bold** *and alignment in* **C3:C4** *to* **Right**.

Note:
Using ranges rather
than name references,
the functions would be:
Covar(A9:A33,B9:B33)
Correl(A9:A33,B9:B33)

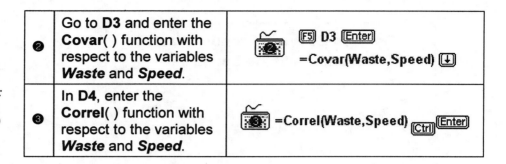

| ❷ | Go to **D3** and enter the **Covar( )** function with respect to the variables **Waste** and **Speed**. | [F5] D3 [Enter] =Covar(Waste,Speed) [↓] |
| ❸ | In **D4**, enter the **Correl( )** function with respect to the variables **Waste** and **Speed**. | =Correl(Waste,Speed) [Ctrl] [Enter] |

Figure 14.4 shows the output.

	A	B	C	D	E
1	**Measuring Relationship Between Waste and Speed**				
2					
3			Covariance	-15.6248	
4		Coefficient of Correlation		-0.73457	
5					

Figure 14.4 Measures of Relation Between *Waste*
and *Speed*

Note that the order of the arguments in both the functions is unimportant. The functions could also be defined as **Covar(Speed,Waste)** and **Correl(Speed,Waste)**. The coefficient of correlation can also be computed as:

=Covar(Speed,Waste)/(StdevP(Speed)*StDevP(Waste))

The value of **r**, the sample correlation coefficient, for *Speed* and *Waste* is **-0.73457**. This indicates that there is a moderately strong relationship between the two variables and that they are related in such a way that when one variable increases, the other decreases. The process team thought that this would be true since **TQM** predicts that as speed and productivity increase, waste decreases.

Exercise 1. Do a correlation analysis for the variables *Waste* and *Delay*. Does there appear to be a correlation there? Does this make sense?

Section 14.7 Finding the Regression Model

The next step in the analysis is finding the regression equation for *Waste* as a function of *Speed*. This will enable the team to predict what waste should be for different values of *Speed*. Such information can be used first to assess the **TQM** program and second to troubleshoot the process, if *Waste* appears to be higher than the predicted value for a given speed. Our model is assumed to be linear. Thus:

$$\text{Waste} = b_0 + b_1\text{Speed}$$

Excel is equipped with functions **Intercept()** and **Slope()** to calculate the b_0 (Y-intercept) and b_1 (line slope) coefficients.

Procedure 14.4 Calculating Coefficients of a Regression Line

Task Description	Mouse/Keyboard
❶ Add two more labels: ◆ In **C5**: **Intercept**, and ◆ In **C6**: **Slope**.	<table><tr><td></td><td>A</td><td>B</td><td>C</td></tr><tr><td>1</td><td colspan="3">Measuring Relationship Betwe</td></tr><tr><td>2</td><td></td><td></td><td></td></tr><tr><td>3</td><td></td><td></td><td>Covariance</td></tr><tr><td>4</td><td colspan="3">Coefficient of Correlation</td></tr><tr><td>5</td><td></td><td></td><td>Intercept</td></tr><tr><td>6</td><td></td><td></td><td>Slope</td></tr></table>
❷ Go to **D5** and enter the **Intercept( )** function with respect to the variables *Waste* and *Speed*.	[F5] D5 [Enter] =Intercept (Waste,Speed) [↓]
❸ In **D6**, enter the **Slope( )** function with respect to the variables *Waste* and *Speed*.	=Slope (Waste,Speed) [Ctrl][Enter]

Note:
Also change the font style of the labels to **Bold** *and alignment in* **C5:C6** *to* **Right**.

Figure 14.5 show the output.

Note, this time the order of the arguments in both the functions is important. The dependent variable (*Waste*) goes before the independent one (*Speed*).

	A	B	C	D	E
1	Measuring Relationship Between Waste and Speed				
2					
3			Covariance	-15.6248	
4		Coefficient of Correlation		-0.73457	
5			Intercept	13.90108	
6			Slope	-0.01289	

=Covar(Waste,Speed)
=Correl(Waste,Speed)
=Intercept(Waste,Speed)
=Slope(Waste,Speed)

**Figure 14.5 Measure of Relation Between Waste and Speed
and Parameters of Their Regression Line**

From the output, we can see that the linear regression equation of **Y** (*Waste*) on **X** (*Speed*) is given by:

Waste = **13.9011 - 0.0129** *Speed*.

Exercise 2. Change the value of *Waste* for the last data point from **7.4** to **12.0**. What happens to the model when there is a value, which appears to be an outlier? What happens to the value of the correlation coefficient?

Section 14.8 Testing Significance of the Model

The method of Least Squares, which is the method used to define the linear regression model, will find the regression equation for any two variables. The fact that an equation is found does not in any way mean that the model is significant.

There are several ways to test whether a regression model is significant:

1 testing whether $\beta_1 \neq 0$;
2 testing whether the correlation coefficient, $\rho \neq 0$; and
3 testing the Regression Sum of Squares vs. Error Sum of Squares.

Note:
*It is not sufficient to look at the **magnitude** of b_1 to make this decision, since the magnitude of the coefficient is related to the order of magnitude of the data.*

Method 1 tests to see if there is a relationship by testing whether the slope coefficient (β_1) is different from 0. You want to test the hypotheses:

$$H_O: \quad \beta_1 = 0$$

$$H_A: \quad \beta_1 \neq 0$$

Procedure 14.5 Performing the Significance of β_1 Test

	Task Description	Mouse/Keyboard
❶	Stay on the **WasteVsSpeed** sheet and invoke the: ♦ **Tools** ♦ **Data Analysis** command.	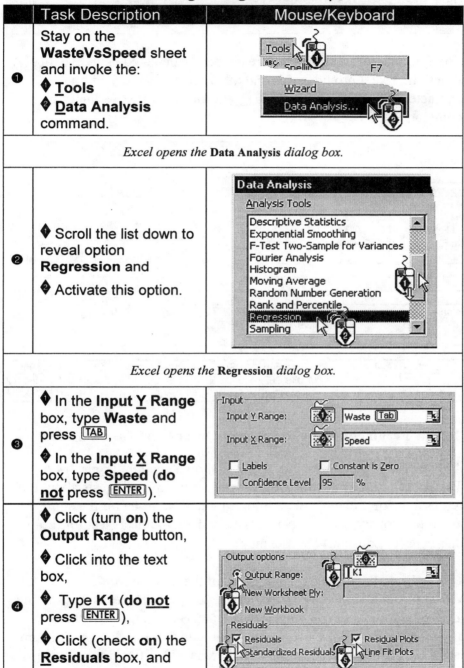
	Excel opens the **Data Analysis** *dialog box.*	
❷	♦ Scroll the list down to reveal option **Regression** and ♦ Activate this option.	
	Excel opens the **Regression** *dialog box.*	
❸	♦ In the **Input Y Range** box, type **Waste** and press (TAB), ♦ In the **Input X Range** box, type **Speed** (**do not** press (ENTER)).	
❹	♦ Click (turn **on**) the **Output Range** button, ♦ Click into the text box, ♦ Type **K1** (**do not** press (ENTER)), ♦ Click (check **on**) the **Residuals** box, and ♦ Click (check **on**) the **Residual Plots** box.	

Note:
K1 *is where you want Excel to generate the* **Regression** *report. Make sure that you have nothing important in the range below and on the right-hand side of* **K1**.

⑤	Click the **OK** command button.	

The estimate of the slope has already been calculated using the **Slope()** function. Excel provides **this** and other regression related characteristics through the **Tools | Data Analysis | Regression** command.

	K	L	M	N	O	P	Q
1	SUMMARY OUTPUT						
2							
3	*Regression Statistics*						
4	Multiple R	0.734569617					
5	R Square	0.539592523					
6	Adjusted R Square	0.519574806					
7	Standard Error	0.432265395					
8	Observations	25					
9							
10	ANOVA						
11		*df*	*SS*	*MS*	*F*	*Significance F*	
12	Regression	1	5.036772443	5.036772	26.95575	2.9017E-05	
13	Residual	23	4.297627557	0.186853			
14	Total	24	9.3344				
15							
16		*Coefficients*	*Standard Error*	*t Stat*	*P-value*	*Lower 95%*	*Upper 95%*
17	Intercept	13.90108351	1.03082622	13.48538	2.08E-12	11.76865991	16.03350712
18	X Variable 1	-0.012894303	0.002483546	-5.19189	2.9E-05	-0.018031901	-0.007756704

Intercept
Slope

Figure 14.6 A Fragment of the Regression Output for *Waste* **vs.** *Speed*

A fragment of the regression output is shown again in Figure 14.6. The bottom part of the output gives the results of two **t** tests about the regression coefficients. In addition to the hypotheses for the slope given previously, you can perform a similar hypothesis test for the intercept (or constant) coefficient β_o.

From the output you can see that the p-value for the slope (**X Variable 1** in **O18**) test is **0.000029 (2.9E-05)**, which means that for any level of significance (α) greater than **0.00003** (which is just about anything) you would reject H_O and conclude that the slope coefficient is not **0**. That is, you conclude that there is a significant relationship between *Waste* and *Speed*.

The output also gives the value of **R Squared** (in **L5**), the **Coefficient of Determination**. The value of **R Squared** can be viewed as the proportion of the variation in the dependent variable **Y** that is explained by or accounted for by the model (its relationship with **X**). For these two variables **R Squared** is about **54%** . Note that the value of **R Squared** is equal to the value of the correlation coefficient squared (r^2).

Section 14.9 Examining the Fit of the Model

Although the statistical test indicates that the regression model is significant, you may wonder just how good the model is at predicting. One way to find this out is to look at the differences between the observed value of **Y** and the predicted value of **Y**. These differences are known as the **residuals**.

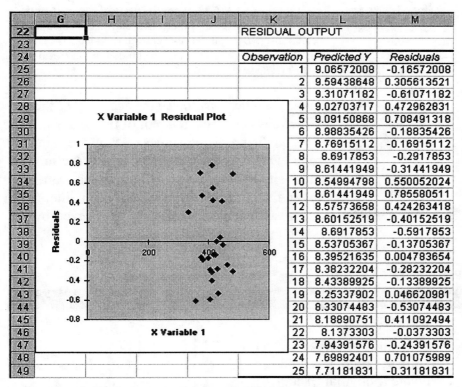

	G	H	I	J	K	L	M
22					RESIDUAL OUTPUT		
23							
24					Observation	Predicted Y	Residuals
25					1	9.06572008	-0.16572008
26					2	9.59438648	0.305613521
27					3	9.31071182	-0.61071182
28					4	9.02703717	0.472962831
29					5	9.09150868	0.708491318
30					6	8.98835426	-0.18835426
31					7	8.76915112	-0.16915112
32					8	8.6917853	-0.2917853
33					9	8.61441949	-0.31441949
34					10	8.54994798	0.550052024
35					11	8.61441949	0.785580511
36					12	8.57573658	0.424263418
37					13	8.60152519	-0.40152519
38					14	8.6917853	-0.5917853
39					15	8.53705367	-0.13705367
40					16	8.39521635	0.004783654
41					17	8.38232204	-0.28232204
42					18	8.43389925	-0.13389925
43					19	8.25337902	0.046620981
44					20	8.33074483	-0.53074483
45					21	8.18890751	0.411092494
46					22	8.1373303	-0.0373303
47					23	7.94391576	-0.24391576
48					24	7.69892401	0.701075989
49					25	7.71181831	-0.31181831

Note:
Your **Residual Plot** *may be found at another location. Check the range on the right-hand side of the* **Regression** *report.*

Figure 14.7 Predicted Values and Residuals of *Waste*

When you were doing the **Tools | Data Analysis | Regression** command, you checked the **Residuals** and **Residuals Plot** boxes. As a result, Excel produced the output as shown in Figure 14.7.

If you look at the values in *Residuals* you will see that the residuals, or errors, range from **-0.6107** (predicted value higher than observed data) to **0.7856** (predicted value lower than observed data).

	A	B	C
8	**Speed**	**Waste**	**Prediction**
9	375	8.9	9.07
10	334	9.9	9.59
11	356	8.7	9.31
12	378	9.5	9.03
13	373	9.8	9.09
14	381	8.8	8.99
15	398	8.6	8.77
16	404	8.4	8.69
17	410	8.3	8.61
18	415	9.1	8.55
19	410	9.4	8.61
20	413	9.0	8.58
21	411	8.2	8.60
22	404	8.1	8.69
23	416	8.4	8.54
24	427	8.4	8.40
25	428	8.1	8.38
26	424	8.3	8.43
27	438	8.3	8.25
28	432	7.8	8.33
29	443	8.6	8.19
30	447	8.1	8.14
31	462	7.7	7.94
32	481	8.4	7.70
33	480	7.4	7.71

Figure 14.8 Model-based Waste Predictions

Exercise 3. Calculate the predicted values of *Waste* using the regression line model ($b_0 + b_1x$) rather than the **Tools | Data Analysis | Regression** command.

Hint:
On the **WasteVsSpeed** sheet, name the **Intercept** cell (**D5**) as **bZero**, the **Slope** cell (**D6**) as **bOne**, and the **Speed** range (**A9:A33**) as **xSpeed** (note that x before **Speed** is intentional—not a typo). In the cell **C8**, type the label **Prediction**. Make the label **Bold** and **Right-aligned**. Select the range **C9:C33**, type **=bZero+bOne*xSpeed** and press CTRL+ENTER. Figure 14.8 shows the calculated predictions. Note that the results were rounded to **0.01**. Compare these values with those included in the **Residual Output** (range **L25:L49**).

Another way to see the accuracy of the predictions is to look at a plot of the actual values of **Y** and the predicted values of **Y** on the same graph. One way to do it is add the **Prediction** range (either **C9:C33** or **L25:L49**) to the scatter plot and then format the new range as a line. The same result can be accomplished, however, by adding the trend line to the scatter plot.

Procedure 14.6 Adding the Tend Line to the Scatter Plot

	Task Description	Mouse/Keyboard
❶	On your **Scatter Plot**, point to one of the data points (note that Excel shows the **Speed** and **Waste** values that make up that point).	

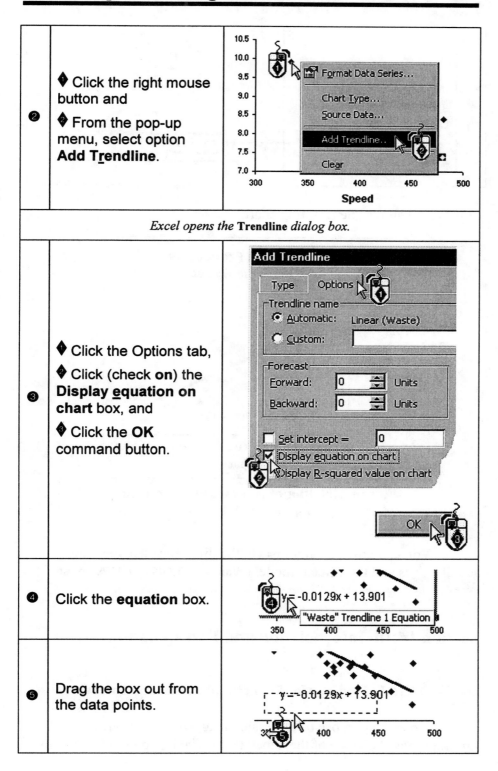

❷	♦ Click the right mouse button and ♦ From the pop-up menu, select option **Add Trendline**.	

Excel opens the **Trendline** *dialog box.*

❸	♦ Click the Options tab, ♦ Click (check **on**) the **Display equation on chart** box, and ♦ Click the **OK** command button.	
❹	Click the **equation** box.	
❺	Drag the box out from the data points.	

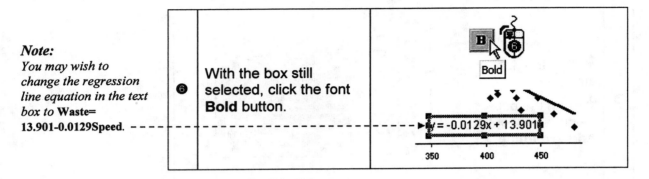

❻ With the box still selected, click the font **Bold** button.

Figure 14.9 shows the resulting chart.

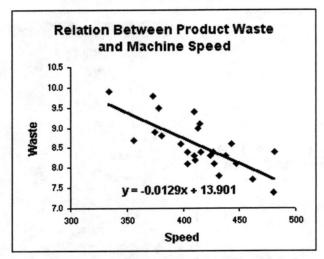

Figure 14.9 Plot of Actual and Predicted Values of Waste

You can now visually inspect the difference between the actual points and their linear model (**Waste = 13.9011–0.0129Speed**).

Section 14.10 Interpolation and Extrapolation

Section 14.10.1 Finding Confidence and Prediction Intervals

Since the purpose of regression analysis is to provide a model for predicting the value of the dependent variable, you might want to look at the predictions for a model and decide how useful they really are.

Excel provides an easy way to obtain the predicted values of **Y** for the selected values of **X**, but determining (1-α)% prediction intervals for individual values of **Y** for a given value of **X** requires an extra work with rather complicated formulas. The (1-α)% lower and upper prediction limits (*LPL* and *UPL*) are defined as

$$b_O + b_1 x \pm t_{\alpha/2} s_{XY} \sqrt{1 + \frac{1}{n} - \frac{(x - \overline{X})^2}{\sum_{k=1}^{n} x_k^2 - n\overline{X}^2}}$$

where

n	the number of observations	=**Count(Speed)**
$t_{\alpha/2}$	= **TDist(α,n-2)**	
s_{XY}	the Standard Error (in Regression Output, =**J7**)	
$\overline{X}$	=**Average(Speed)**	
$\sum_{k=1}^{n} x_k^2$	=**SumSq(Speed)**, the sum of squares of *Speed* values	

You may wish to develop the formulas all by yourself, or you can do it with a help of the **PredictionIntervalLimits** macro. This macro requires that two columns of the independent and dependent data, **X** and **Y** (in this case—*Speed* and *Waste*), be stored on an empty sheet starting from cell **A1**.

Procedure 14.7 Generating Prediction Intervals

	Task Description	Mouse/Keyboard
❶	From the name box, select the name **Speed**.	
	*Excel switches to the **Data** sheet and selects the **Speed** range.*	

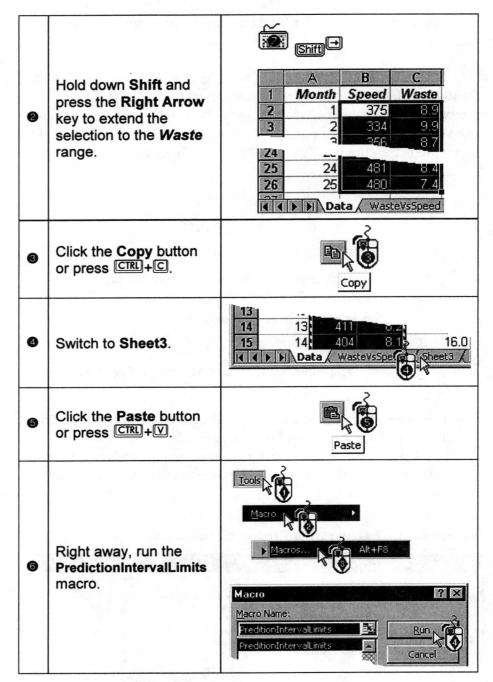

| ❷ | Hold down **Shift** and press the **Right Arrow** key to extend the selection to the *Waste* range. | |
| ❸ | Click the **Copy** button or press CTRL+C. | |

Note:
If Sheet3 *is not free, use another one. You can also insert a new sheet.*

❹	Switch to **Sheet3**.	
❺	Click the **Paste** button or press CTRL+V.	
❻	Right away, run the **PredictionIntervalLimits** macro.	

A fragment of the output is shown in Figure 14.10.

	A	B	C	D	E	F
1			Prediction Interval Limits			
2						
3		b_0	13.90108351	xBar	413.6	
4		b_1	-0.012894303	t_α	2.068654794	
5		α	5%	SumX^2	4306918	
6		n	25	s_{XY}	0.432265395	
7						
8						
9	X	Y	Prediction	Residual	LPL	UPL
10	375	8.9	9.0657	-0.1657	8.1325	9.9990
11	334	9.9	9.5944	0.3056	8.5950	10.5938
		9.7	9.3107	-0.6107	8.3520	10.2694
31	447	8.1				
32	462	7.7	7.9439	-0.2439	6.9987	8.8891
33	481	8.4	7.6989	0.7011	6.7235	8.6744
34	480	7.4	7.7118	-0.3118	6.7382	8.6855

Figure 14.10 Prediction Interval Limits for Waste vs. Speed at 95% Confidence

Suppose you want to obtain the predicted values and limits of *Waste* for a speed, say **400** PPM, which is not in the data set. In order to do this, copy the formulas for *Prediction*, *LPL*, and *UPL*, from cells **C10**, **E10**, and **F10**, and paste them onto cells **C8**, **E8**, and **F8**. Then enter **400** into cell **A8**. You should get a result as shown in Figure 14.11.

	A	B	C	D	E	F
1			Prediction Interval Limits			
2						
3		b_0	13.90108351	xBar	413.6	
4		b_1	-0.012894303	t_α	2.068654794	
5		α	5%	SumX^2	4306918	
6		n	25	s_{XY}	0.432265395	
7						
8	400		8.7434		7.8288	9.6580
9	X	Y	Prediction	Residual	LPL	UPL
10	375	8.9	9.0657	-0.1657	8.1325	9.9990

Figure 14.11 Waste Prediction and Limits for Speed = 400

Note:
*If you want prediction and limits of another value of the independent variable, enter it into cell **A8**. To see the prediction limits for another significance level, enter the level into cell **C5**.*

From the output you see that for a *Speed* of **400** the predicted *Waste* is **8.7434%**. The values for the prediction interval range from **7.8288%** to **9.6580%**. This tells the team that, if waste is measured when the machine is run at **400** PPM the individual value of *Waste* will range from **7.83%** to **9.66%**. The prediction intervals for a regression model give information on how *useful* the model is. If the intervals are too wide, because the relationship is not that strong, then the range of the intervals will be too wide to be meaningful or useful to the people who will use them. For example, if the prediction interval for a speed of **400** PPM were to range from **3.05%** to **8.04%**, the manufacturing team would not find this useful in planning.

Exercise 4. Obtain the prediction limits for a confidence level of **99%**. What happens to the prediction and confidence intervals? What happens if you change it to **90%**?

Using the model to predict values of **Y** is valid as long as the values you choose for **X** are within the range of the original data. That is, as long as you are **interpolating**. You might be interested in predicting *Waste* for speeds above **350** or below **480**. This is called **extrapolation** and is **not advisable**. The model is only valid over the range of *Speed* observed!

Section 14.10.2 Plotting Prediction Intervals

It is interesting to look at the relationship among the regression line and the prediction intervals. Figure 14.12 shows a scatter chart for **Y** (*Waste*), **Prediction**, **LPL**, and **UPL** vs. **X** (*Speed*).

Exercise 5. Create a chart showing the actual Waste, predicted Waste and the Upper and Lower Limits.

Hint:
To create this chart, select ranges **A9:C34**, **E9:F34** and invoke the **Chart Wizard**. Choose the **XY Scatter Chart** type. Eliminate **Gridlines** and **Legend**. Set the **Plot Area Color** and **Borders** to **None**. Change the Patterns for **Prediction** to **Line** only and for the limits (**LPL**, **UPL**) to both **Line** and **Markers**. Change the **X Minimum** to **300** , **Y Minimum** to **6** and **Y Maximum** to **11**.

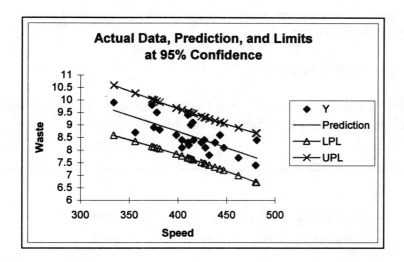

Figure 14.12 Scatter Plot of Waste vs. Speed

From the output you can see the relationship between the prediction line and prediction interval limits. Looking at the prediction interval limits you can see that they are slightly better (tighter) at the center and get wider as you move away from the center.

Note:
The X mean and Y mean point $(\bar{x}, \bar{y})$ *defines the center of the regression line.*

Section 14.11 Investigative Exercises

In the following exercises you are asked to use the skills introduced in the previous chapters to extract information from the tissue strength data file. You are provided with space to answer the questions and paste in graphical output from the program. If you do not have access to a printer, you can sketch the graphs in the spaces provided.

1. a) Create a scatter plot for *Speed* and *Delay*.

b) Does there appear to be a relationship between *Speed* and *Delay*? Describe the relationship.

c) Find the correlation coefficient for *Speed* and *Delay*. Does its value agree with your description of the relationship?

2. a) Find the regression model for *Delay* and *Speed*. Write the regression equation.

b) Is the regression significant?

c) Plot the regression line along with the data Does the plot indicate that the model does a good job of predicting *Delay?*

---

3. a) Use the model to predict *Delay* for speeds of **380** and **470**. What do you learn from the prediction and limits?

b) Use the model to predict *Delay* for a speed of **600** PPM. Does the answer make sense? Do you think it is valid? Why or why not?

---

4. Use Excel to perform a complete regression analysis for the variables *Speed* and *RateOper*. Is the model a good one?

---

5. Use the model you just found to predict the value of *RateOper* for a speed of **400**. Do you think that the prediction and interval limits would be useful to the company?

6.a) Make a scatter plot of *Speed* and *Quality*. Do you think that there is a relationship between the two?

b) Perform a correlation analysis for the two variables. Does it indicate that there is a strong relationship?

7. Find the regression model for *Speed* and *Quality*.

---

8. Change the last **10** values for *Quality* to .97. Redo the regression analysis. Describe the results and how they compare to the model you found in the previous exercise.

9. The manufacturing company decides that it wants to know if there has been a steady increase in *RateOper* in the months since the TQM program was instituted.

a) How might they accomplish this?

b) Plot *RateOper* against *Month*. Does it appear that there has been a steady increase?

c) Find a regression equation that predicts *RateOper* as a function of *Month*. What is the model?

d) Is it significant?

e) Find the residuals. Does the fit of actual vs. predicted appear to be good?

---

10. Prepare a report for management telling them about the effectiveness of the **TQM** program. Perform any additional analyses that you think you need to make a complete report. Include any plots that are relevant.

Chapter 15 "Do Your Tissues Rip?"

Pulling It All Together

Section 15.1 Overview

Statistical Objectives: There are no new statistical concepts to be learned in this chapter. The chapter is designed to allow you to use what you have learned through the exercises in this workbook. You will be asked to analyze a large data set but this time you will not be given any help. You will need to use statistical tools from several of the preceding chapters and it will be up to you to decide which tools are most appropriate. You are called upon to "pull it all together" !

Section 15.2 Problem Statement

Have you ever pulled the first tissue out of the box and had it tear? Have you ever opened a box of tissues and in trying to get one tissue out, ended up with several tissues? The problem causing both of these "nasty" things to happen is that there is not enough airspace in the box. This problem became an important issue for a large manufacturer of tissues when an unusually high number of complaints were registered.

Airspace is defined to be the amount of space between the top of the tissues and the top of the box. It is measured in millimeters and it should be at least 9 mm. Even if there is 9 mm of airspace when the box is manufactured, there may not be enough airspace by the time the customer opens the box. This is due to a phenomenon which is called "growback". As the box sits in the warehouse or on the supermarket shelf the tissues, which have been heavily compressed when they were put into the box, begin to expand or "growback". Thus the airspace is reduced.

In order to understand this data set you need to know a little bit about how a box of tissues is made. In Figure 15.1 you see a picture of a hardroll. Each box of **250** tissues is made from **25** hardrolls each of which has different slit positions as shown in the diagram.

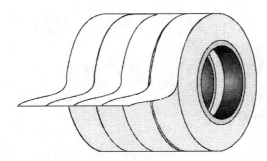

Figure 15.1 Schematic of a Hardroll

One tissue is taken from each slit and they are pressed into a tissue box and sealed.

An experiment was designed to investigate the hypothesis that the mean airspace differs by position in the hardroll from which the tissues were made. This hypothesis was proposed because of two facts:

1. As the hardroll sits in the warehouse, the outside collects moisture. Thus boxes made from tissues taken from the outside of the roll might have less "growback".

2. The core (very center) of the roll is very compressed (a hardroll weighs **1000** pounds). Thus boxes of tissues made from the core have the "life stretched out of them".

The data was collected in the following fashion:

- Sample **5** times through the hardroll. Since a hardroll lasts for **4** hours the **5** observations were taken at the beginning and once an hour thereafter. This resulted in observations at the top, the core and **3** in the middle.

- Each sample consisted of a case of tissues (**24** cartons).

- A total of **4** cases were sampled every hour - one was used to obtain in-process measurements. The other **3** cases were saved for observations taken after **24** hours, **2** weeks and **4** weeks.

Section 15.3 Characteristics of the Data Set

FILENAME: Ch15Dat.xls   An Excel Workbook
SIZE: COLUMN   3
             ROWS   480

A fragment of the actual data file are shown in Figure 15.2.

	A	B	C	D	E	F
1	**Position**	**Time**	**Airspace**			
2	1	1	23			
3	1	1	25			
4	1	1	23			
		1	23			
478	5					
479	5	4	15			
480	5	4	16			
481	5	4	15			

Data / Sheet2 / Sheet3

Figure 15.2 A Fragment of the Chapter 15 Data Set

Notes on the data set:

1. The variable *Position* indicates from what part of the hardroll the sample was taken. The values for position are 1,2,3,4, or 5.   Position 1 indicates it was taken from the outside of the roll and position 5 indicates it was taken from the core.

2. The variable *Time* indicates when the measurement was taken: 1= In-process, 2=after 24 hours, 3=after 2 weeks   4= after 4 weeks.

3. The variable *Airspace* is the measurement of the airspace measured in mm.

Open

Open the data set named **Ch15Dat.xls** from your data disk. Note that Procedure 3.1, on page 33, contains detail instruction about how to open an Excel workbook.

Click the Open *button or use the* File | Open *command (*ALT+F, O*) to open the file.*

Section 15.4 Investigative Exercises

Analyze the data using whatever techniques you feel are appropriate. Be sure to explain why you have selected a particular technique. Your analysis must address the following questions:

1. Is there a difference in airspace due to position in the hardroll?

2. Is there a difference in airspace by time tested?

3. How can you estimate "growback"?

Note:

If you also open the **MacDoIt.xls** workbook, you will have all macros presented in the preceding chapters available.

Notes

Notes

Notes

Notes

Notes

Notes

Notes